Health Promoting Compounds in Milk and Dairy Products

Health Promoting Compounds in Milk and Dairy Products

Guest Editors

Pamela Manzi
Mena Ritota

MDPI

Basel • Beijing • Wuhan • Barcelona • Belgrade • Novi Sad • Cluj • Manchester

Guest Editors

Pamela Manzi
CREA—Centro di ricerca
Alimenti e Nutrizione
Rome
Italy

Mena Ritota
CREA—Research Centre for
Food and Nutrition
Rome
Italy

Editorial Office
MDPI AG
Grosspeteranlage 5
4052 Basel, Switzerland

This is a reprint of the Special Issue, published open access by the journal *Molecules* (ISSN 1420-3049), freely accessible at: https://www.mdpi.com/journal/molecules/special_issues/milk_dairy_products_compound.

For citation purposes, cite each article independently as indicated on the article page online and as indicated below:

Lastname, A.A.; Lastname, B.B. Article Title. *Journal Name* **Year**, *Volume Number*, Page Range.

ISBN 978-3-7258-7352-4 (Hbk)
ISBN 978-3-7258-7353-1 (PDF)
https://doi.org/10.3390/books978-3-7258-7353-1

Contents

About the Editors

Pamela Manzi

Pamela Manzi is a Senior Researcher at the CREA Council for Agricultural Research and Economics—Research Centre for Food and Nutrition in Rome, where she has worked since 1996. Her scientific interests focus on the characterization of food products, with particular expertise in the nutritional and functional quality of milk and dairy products, the identification of quality indicators, and the study of correlations between food composition, technological process, and consumer information. She has served as Principal Investigator for the PRIMA project "VEGGIE-MED-CHEESES," coordinated operative units in several Italian Ministry of Agriculture-funded projects, and contributed as Scientific Manager to regional initiatives on traditional agri-food products. Her achievements include a long record of international and national research leadership, participation in the management committee of the EU School Fruit, Vegetables and Milk Scheme, and the publication of extensive peer-reviewed work covering topics from dairy chemistry to functional components of food. She also contributes to academic training by teaching on the PhD programme in Food Science and the Master's degree programme in Human Nutrition Science at the University of Rome.

Mena Ritota

Mena Ritota has been a researcher at CREA Food and Nutrition since 2015. She has a PhD in Chemistry (Analytical Methodologies curriculum), and is a member of the hub "Sustainable Food Design: pathways of sustainability toward ecological transition," which aims to promote a new paradigm of sustainable production through research projects, models for the redesign/reengineering of Italian food, training and scientific outreach activities, and the creation of a national stakeholder network. Her current research focuses on: analytical characterization of food biodiversity (quality and traceability) of agri-food products of both animal and plant origin; the study of the chemical-nutritional quality of foods, with particular attention to functional and/or nutritionally relevant molecules and the development of associated analytical methods; the variation in the chemical–nutritional profile of foods depending on production/cultivation/processing technologies; and the chemical, nutritional and technological properties of edible mushrooms and their promotion as models of circular economy.

molecules

Editorial

Health Promoting Compounds in Milk and Dairy Products

Mena Ritota * and Pamela Manzi

Consiglio per la Ricerca in Agricoltura e l'Analisi dell'Economia Agraria, Centro di Ricerca Alimenti e Nutrizione, Via Ardeatina 546, 00178 Rome, Italy; pamela.manzi@crea.gov.it
* Correspondence: mena.ritota@crea.gov.it

1. Introduction

Milk represents a fundamental component of the human diet, as it supplies most of the nutrients essential for growth and development. It can be consumed in its natural or fermented form or processed into yoghurt, cheese, and a wide variety of other dairy products; it is also the basic ingredient for many products. Milk and dairy products are widely recognized as rich sources of calcium, which is present in a highly bioavailable form [1], but they also contribute significant amounts of specific minerals and vitamins [2].

The complex composition of milk, characterized by a wide range of nutrients and bioactive compounds and by the interactions among these components, positively influences digestion, nutrient absorption, and overall human health. This holistic perspective led the International Dairy Federation (IDF) in 2019 to introduce the concept of the "dairy matrix" [3]. Scientific evidence supporting this concept demonstrates that the health benefits of dairy products arise from the entire food matrix rather than from individual components alone. This dairy matrix exerts a wide range of effects, including antimicrobial, biostatic, antihypertensive, ACE-inhibitory, anticarcinogenic, anti-obesity, immunomodulatory, antidiabetic, cholesterol-lowering, probiotic, and prebiotic actions [4]. Consequently, the consumption of milk and dairy products goes beyond providing basic nutrition, as milk components exert relevant metabolic effects and contribute significantly to overall health.

The overall quality of milk and dairy products, as well as their content of health-promoting compounds, is affected by multiple aspects, including genetic factors [5], farming and feeding systems [6], technological processes (e.g., heat treatments [7], fermentation [8], or the addition of specific ingredients rich in bioactive compounds in the case of new product formulations [9], etc. For this reason, in recent years, emerging technologies—e.g., nanotechnology [10] and novel thermal and non-thermal processing [11]—have also been developed to provide new solutions for the incorporation of bioactive compounds into dairy products, enhancing the functional properties of these food items without compromising consumers' sensory perception.

This Special Issue, entitled "Health-Promoting Compounds in Milk and Dairy Products", was conceived to achieve the following aims: (i) explore recent advances in analytical methods for the assessment of bioactive compounds in dairy products; (ii) update current knowledge on the nutritional evaluation of milk-based foods; (iii) collect recent studies aimed at improving the nutritional value of milk and dairy products; and (iv) examine the effects of bioactive compounds from dairy products within the human diet.

Following a rigorous peer-review process, eight papers were accepted, including six original research articles, one review, and one opinion paper.

Molecules **2026**, *31*, 919 https://doi.org/10.3390/molecules31060919

2. An Overview of the Published Articles

Lipids are key components of milk from both a nutritional and economic perspective. They supply a significant amount of energy and contribute to the sensory and physical properties of dairy products. In addition, milk fat acts as a vehicle for naturally occurring fat-soluble vitamins (A, D, E, and K) and β-carotene, a precursor of vitamin A [Contribution 1]. Milk fat is also naturally rich in saturated fatty acids (SFAs), and for this reason dairy fat is often perceived by consumers as increasing the risk of cardiovascular disease, metabolic syndrome, or obesity, leading to a preference for fat-free dairy products. Nonetheless, this negative perception is not supported by the most recent scientific evidence which suggests that dairy products could promote human health thanks to the presence of certain bioactive fatty acids [12]. For these reasons, accurate determination of the fatty acid profile of milk is of considerable importance. A range of analytical techniques is currently available for fatty acid analysis in food matrices, including high performance liquid chromatography (HPLC), gas chromatography (GC), near-infrared and mid-infrared spectroscopy (NIRS and MIRS, respectively), etc. Although conventional chemical methods such as HPLC and GC provide high analytical accuracy, they are often time-consuming, labor-intensive, and costly. In contrast, infrared spectroscopy-based approaches offer rapid and low-cost alternatives for predicting milk fatty acid composition [13], although further optimization is still required to enhance their predictive performance. Within this context, Zhao et al. [Contribution 1] evaluated the feasibility of using mid-infrared spectroscopy to predict the fatty acid composition of milk from Chinese Holstein cattle. The authors conducted a comprehensive comparison of four regression models, five spectral pre-processing methods, two spectral ranges, and two units of fatty acid expression (g/100 g milk and g/100 g fat). Among the tested approaches, random forest regression combined with derivative- and Savitzky–Golay-based preprocessing achieved the highest prediction accuracy. Seventeen fatty acids and fatty acid groups showed reliable predictive performance ($R^2 \geq 0.75$ and RPD ≥ 2), particularly when expressed on a milk basis. The study of Zhao et al. [Contribution 1] represents the first systematic MIRS-based evaluation of milk fatty acids in Chinese Holstein cows and highlights the potential of MIRS as a high-throughput, cost-effective phenotyping tool for genetic evaluation and dairy breeding programs.

Milk is widely recognized for the high nutritional quality of its protein, with caseins accounting for approximately 80% of the total protein fraction. Caseins are a heterogeneous group of proteins consisting of four polymorphisms: αS1-casein, the most abundant (38%), followed by β-casein (36%), κ-casein (13%) and αS2-casein (10%) [14]. β-caseins occur in several genetic variants, among which β-casein A1 and A2, differing by a single amino acid substitution at position 67, are the most common in European cattle breeds [15]. This structural difference is particularly relevant because β-casein A1 can release β-casomorphin-7 (BCM-7) during digestion, a bioactive peptide that has been associated with potential adverse health effects, including gastrointestinal disorders, type 1 diabetes, and coronary heart disease [16,17]. The health concerns related to β-casein A1 have led to increasing consumer demand for milk produced exclusively from A2A2 cows, which is marketed as a functional food and typically commands a premium price. Consequently, there is a strong need for analytical methods that are rapid, reliable, and sensitive enough to authenticate A2A2 milk and detect potential adulteration. As already mentioned above, conventional analytical techniques, such as Liquid Chromatography (LC), Enzyme Linked Immunosorbent Assay (ELISA), Polymerase Chain Reaction (PCR), Polyacrylamide Gel Electrophoresis (PAGE), restriction fragment length polymorphism (RFLP)-PCR, and Fourier-transform infrared (FTIR) spectroscopy, are often time-consuming, expensive, labor-intensive, or insufficiently sensitive, and many of them are unable to simultaneously distinguish and quantify β-casein A1 and A2 at the protein level [Contribution 2]. To address these limitations, Elferink

https://doi.org/10.3390/molecules31060919

et al. [Contribution 2] developed a competitive microsphere-based multiplex immunoassay capable of simultaneously detecting and distinguishing β-casein A1 and A2 in raw and pasteurized milk samples. Through optimized buffer conditions, the assay achieves <10% cross-reactivity between βCA1 and βCA2, demonstrating exceptional antibody specificity for nearly identical proteins. The assay detected adulteration of pure A2A2 milk with common bovine milk at levels as low as 1%, meeting and exceeding relevant regulatory sensitivity expectations. Beyond fraud detection, the method developed by the authors [Contribution 2] enables cow milk phenotyping directly from milk, providing a faster and less invasive alternative to DNA-based genotyping. Furthermore, compared to ELISA, PCR, and FTIR-based methods, the developed method is faster, simpler, has a semi-high throughput, and is well suited for routine quality control.

Regarding the mineral fraction, dairy products are widely recognized as excellent sources of calcium. However, the study by Manzi et al. [Contribution 3] on selected Italian cheeses showed that these products can also represent a valuable dietary source of zinc, an essential trace element involved in numerous catalytic, structural, and regulatory functions in the human body. The authors analyzed a total of 57 PDO (Protected Denomination of Origin) and Traditional Italian cheeses, reporting zinc concentrations ranging from 0.39 to 7.75 mg per 100 g of cheese. The considerable variability in zinc content observed among the samples was attributed to several factors, including feeding system and animal species (cow, sheep, goat, and buffalo), but it was mainly associated with the cheese-making process. Curdling and salting generally represent the two processing stages mainly affecting the mineral composition of cheese [18]. In addition, the authors [Contribution 3] assessed the nutritional relevance of zinc in the analyzed cheeses, showing that approximately half of the samples provided more than 20% of the European Daily Reference Intake for zinc (10 mg/day) [19] with a single serving of cheese (50 or 100 g), whereas only two samples contributed less than 4%, due to the acidic rather than enzymatic cheesemaking process. Overall, these findings support the role of cheese consumption as a valuable means of supplying dietary zinc.

Among fat-soluble vitamins, vitamin D plays a crucial role in the regulation of calcium metabolism and the maintenance of bone health [20,21]. Recent evidence also highlights its involvement in modulating immune system function [22,23]. In the absence of fortified foods, the contribution of the diet to the daily amount of vitamin D is generally scarce, and deficiencies of this vitamin have been reported in various population groups, including individuals living in regions with high sun exposure. Although the natural vitamin D content in cow's milk is relatively low— typically ranging from 0.1 to 1 μg/L in whole milk [24,25]—vitamin D fortification of milk has proven to be an effective strategy for increasing population intake in many countries, such as in Canada, Finland, Sweden and Austria [26]. As a result, this practice is currently mandatory in some countries and strongly recommended in others [26]. In this context, Contribution 4 provides a comprehensive review of current evidence on vitamin D's role in health and disease and explores how milk's structural properties (particularly those related to the milk fat globule membrane and casein micelle organization), as well as industrial processing and storage conditions, impact vitamin D delivery. The authors emphasize the suitability of milk as an efficient vehicle for vitamin D, capable of supplying a modest yet nutritionally meaningful amount that contributes to achieving the adequate daily intake for this vitamin.

In the dairy sector, small-scale producers are increasingly focusing on innovation and high-quality products to meet consumer demand and enhance their market position. One common approach is the enrichment of cheeses with additional ingredients, such as colorants, spices, and flavoring agents, which can enhance sensorial, physical, chemical and nutritional quality of the final product. Saffron serves as a notable example of this practice.

https://doi.org/10.3390/molecules31060919

It is typically added into cheeses as a natural coloring agent but also contributes bioactive properties, due to the presence of crocins (water-soluble carotenoids responsible for saffron color and antioxidant activity), picrocrocin (responsible for saffron bitterness), and safranal (responsible for aroma and antioxidant potential). Ritota et al. [Contribution 5] were the first to evaluate saffron bioactive compounds in cheese by using a single UHPLC method and reported the first UHPLC-based assessment of safranal in cheese. Crocins were detected in all samples, with trans-4-GG identified as the predominant form, whereas picrocrocin was not detected, likely due to its degradation into safranal during cheesemaking. Safranal was found only in one ewe cheese with a high saffron content, probably because of its volatility, which makes it challenging to analyze accurately using liquid chromatography. Additionally, Contribution 5 assessed for the first time the antiproliferative effects of saffron within a food matrix, whereas prior studies had focused mainly on pure compounds (mostly crocin) or saffron extracts. The results showed no significant effect on CaCo2 cells (colon cancer model), while significant antiproliferative effects were observed on HeLa (cervix carcinoma) and MDA-MB-231 (breast cancer) cells. According to the authors [Contribution 5], the chemical environment of the cheese modulates the activity of saffron bioactive compounds, which is influenced not only by crocin concentration but also by residual protein/peptides and fat contents in the extracts. These findings support the concept of a "food matrix effect", where the biological activity of compounds is modified by interactions within the entire food system. They also provide preliminary evidence that saffron-enriched cheeses may have functional food potential, thus opening opportunities for product innovation, added value for producers, and the development of functional dairy products.

The enrichment of dairy products with bioactive compounds was also investigated by Znamirowska et al. [Contribution 6], who added different sources of vitamin C (rosehip, acerola and ascorbic acid in powder form) to probiotic fermented milk to enhance its nutritional value and functionality. Compounds with antioxidant activity (phenolics and flavonoids) in rosehip and acerola were found to better protect vitamin C compared to pure ascorbic acid (in the synthetic form), highlighting a clear food matrix effect on vitamin C stability. The source of vitamin C (natural fruit matrix vs. synthetic) also had a significant impact on probiotic viability, product stability, texture, and sensory quality, with natural vitamin C sources enhancing overall sensory perception of the fermented milk. After 21 days of storage, *Lactobacillus rhamnosus* counts remained above 8 log CFU/g in all samples, exceeding the therapeutic minimum and confirming that all designed products met probiotic standards. Given the advantages of natural products and positive consumer perception, the authors [Contribution 6] suggest that rosehip and acerola are promising candidates for the development of functional dairy products for the market. This Contribution provides results with direct industrial applicability, offering practical guidance on formulation limits (maximum vitamin C dose without causing protein denaturation), effects on gel structure, storage stability data, and sensory acceptability confirmation.

Czarnowska-Kujawska and Paszczyk [Contribution 7] proposed the natural folate bio-fortification of milk through fermentation as a promising alternative to synthetic folic acid fortification, a common practice that can have drawbacks (e.g., masking vitamin B12 deficiency, potential risks of excessive intake). Fermented milk represents an ideal matrix for bio-fortification because it contains folate-binding proteins that enhance stability and bioavailability, is widely consumed, and can be enriched in situ through microbial metabolism. The authors [Contribution 7] evaluated folate production in fermented milk using different commercially available starter cultures, including *Streptococcus thermophilus*, *Lactobacillus delbrueckii* subsp. *Bulgaricus*, *Lactococcus lactis*, and *Bifidobacterium bifidum*. In addition to folate synthesis, these probiotic bacteria were assessed for their ability to

https://doi.org/10.3390/molecules31060919

synthesize conjugated linolenic acid *cis*9*trans*11 C18:2 (CLA). The results showed that mixed starter cultures were more effective than single cultures in producing folate, with the combination of *Lactobacillus* spp., *Streptococcus thermophilus*, and *Bifidobacterium bifidum* yielding the highest folate content (≈105 µg/kg) in fermented milk. Compared to pasteurized milk, folate levels increased 1.6–2.8 times in fermented products. However, microbial bio-fortification alone was insufficient to meet recommended daily folate intake, as the highest folate content would only cover approximately 25% of daily requirements if 1 kg of product were consumed. Furthermore, the authors highlighted that folate stability during storage was significantly influenced by starter culture composition, storage time, and the acidic environment, with losses reaching up to 65% in some formulations. Starter culture selection also impacted on the fatty acid profile, with fermented milks containing *Bifidobacterium bifidum* showing significant increases in CLA during storage, and CLA levels, like folate content, being affected by storage time. Contribution 7 represents one of the few studies combining vitamin bio-fortification and lipid functional enhancement in the same fermented dairy system. The findings strengthen the concept that functional dairy products should be designed through careful microbial consortia selection to maximize multiple health-promoting compounds while maintaining stability during shelf life but also highlight the need for further research to optimize conditions for producing effective functional dairy products.

Dabo et al. [Contribution 8] provided a comprehensive and critical overview of recent advances in whipping cream formulations, an oil-in-water emulsion widely used in the food industry and increasingly popular in the consumer market. The authors examined how structural and compositional parameters affect the performance, efficiency, and biological or chemical activity of whipping creams, while also addressing the main limitations of current approaches, including challenges related to stability, scalability, cost, and practical implementation. This analysis highlights the gap between laboratory-scale research and industrial or clinical translation.

In response to increasing consumer demand for healthier and higher-quality options, current and future research on whipping cream is focusing on the development of healthier products, including low-fat, low-sugar, functional whipping creams, and formulations enriched with probiotics. Another major trend is the development of plant-based alternatives to traditional dairy whipping cream. However, achieving sensory and functional properties comparable to conventional dairy products remains challenging, as plant proteins often require structural modifications (hydrolysis, particle formation, or microgel creation) and synergistic interactions with emulsifiers to replicate the performance of dairy proteins. Furthermore, a deeper understanding of the mechanisms governing low-molecular-weight surfactants and fat nucleation processes is essential to improve control over whipping cream structure and quality. Overall, Contribution 8 offers a valuable resource for researchers by consolidating dispersed knowledge, providing critical analysis, and presenting a forward-looking roadmap for future developments in the field.

3. Conclusions and Future Perspectives

This Special Issue highlights the central role of milk and dairy products as complex food systems in which nutritional value and health effects arise from the interaction among multiple components rather than from single nutrients alone. The concept of the dairy matrix clearly emerges across the collected contributions, reinforcing the idea that the structural organization of milk components (lipids, proteins, minerals, vitamins, and bioactive compounds) plays a crucial role in modulating digestion, bioavailability, and biological activity.

https://doi.org/10.3390/molecules31060919

The published studies also reveal significant progress in both analytical and technological approaches. The development of rapid, reliable, and cost-effective analytical tools, such as mid-infrared spectroscopy and multiplex immunoassays, offers promising opportunities for high-throughput quality control, product authentication, and on-farm phenotyping, supporting innovation along the entire dairy value chain. At the same time, updated evidence on the mineral and vitamin composition of dairy products confirms their significant contribution to micronutrient intake, while simultaneously emphasizing how processing and formulation can enhance or limit their nutritional impact.

A recurring theme across several contributions is the enrichment and bio-fortification of dairy products with health-promoting compounds, achieved through the incorporation of natural ingredients, targeted fermentation strategies, or optimized fortification practices. These studies provide clear evidence that the food matrix plays a decisive role in determining compound stability, bioactivity, probiotic viability, and sensory acceptance. It is worth underlining that the development of functional dairy products must be approached holistically. This approach should balance nutritional improvement with considerations of technological feasibility, shelf-life stability, and consumer perception.

Looking forward, future research should focus on a deeper understanding of matrix–compound interactions, particularly during digestion and metabolism, in order to improve the prediction of health-related outcomes. Key research directions include the optimization of microbial consortia for multi-compound bio-fortification, the further development of non-thermal and emerging processing technologies, and the validation of functional effects through well-designed in vivo and clinical studies. Furthermore, bridging the gap between laboratory-scale innovation and industrial application remains a critical challenge in the design of new functional dairy products.

Overall, the contributions collected in this Special Issue confirm that milk and dairy products continue to be a dynamic and highly promising system for nutritional innovation. By combining advanced analytics, smart processing, and a matrix-oriented perspective, future developments in the dairy sector can lead to products that not only meet consumer expectations but also contribute meaningfully to public health and sustainable food systems.

The guest editors sincerely thank all the authors for the high quality of their contributions and extend their sincere appreciation to the reviewers for their insightful and constructive assessments. They also thank the *Molecules* editorial team for their professional guidance and support throughout the publication process. Hopefully, this Special Issue will serve as a valuable reference point and inspire further research into the field of health-promoting compounds in milk and dairy products.

Author Contributions: Conceptualization, M.R. and P.M.; writing—original draft preparation, M.R.; writing—review and editing, M.R. and P.M. All authors have read and agreed to the published version of the manuscript.

Funding: This research received no external funding.

Data Availability Statement: No new data were created or analyzed in this study. Data sharing is not applicable to this article.

Conflicts of Interest: The authors declare no conflicts of interest.

List of Contributions:

1. Zhao, X.; Song, Y.; Zhang, Y.; Cai, G.; Xue, G.; Liu, Y.; Chen, K.; Zhang, F.; Wang, K.; Zhang, M.; et al. Predictions of Milk Fatty Acid Contents by Mid-Infrared Spectroscopy in Chinese Holstein Cows. *Molecules* **2023**, *28*, 666. https://doi.org/10.3390/molecules28020666.

https://doi.org/10.3390/molecules31060919

2. Elferink, A.J.W.; Entiriwaa, D.; Bulgarelli, P.; Smits, N.G.E.; Peters, J. Development of a Microsphere-Based Immunoassay Authenticating A2 Milk and Species Purity in the Milk Production Chain. *Molecules* **2022**, *27*, 3199. https://doi.org/10.3390/molecules27103199.
3. Manzi, P.; Di Costanzo, M.G.; Ritota, M. Content and Nutritional Evaluation of Zinc in PDO and Traditional Italian Cheeses. *Molecules* **2021**, *26*, 6300. https://doi.org/10.3390/molecules26206300.
4. Pellegrino, L.; Marangoni, F.; Muscogiuri, G.; D'Incecco, P.; Duval, G.T.; Annweiler, C.; Colao, A. Vitamin D Fortification of Consumption Cow's Milk: Health, Nutritional and Technological Aspects. A Multidisciplinary Lecture of the Recent Scientific Evidence. *Molecules* **2021**, *26*, 5289. https://doi.org/10.3390/molecules26175289.
5. Ritota, M.; Comitato, R.; Manzi, P. Cow and Ewe Cheeses Made with Saffron: Characterization of Bioactive Compounds and Their Antiproliferative Effect in Cervical Adenocarcinoma (HeLa) and Breast Cancer (MDA-MB-231) Cells. *Molecules* **2022**, *27*, 1995. https://doi.org/10.3390/molecules27061995.
6. Znamirowska, A.; Szajnar, K.; Pawlos, M. Effect of Vitamin C Source on Its Stability during Storage and the Properties of Milk Fermented by Lactobacillus rhamnose's. *Molecules* **2021**, *26*, 6187. https://doi.org/10.3390/molecules26206187.
7. Czarnowska-Kujawska, M.; Paszczyk, B. Changes in the Folate Content and Fatty Acid Profile in Fermented Milk Produced with Different Starter Cultures during Storage. *Molecules* **2021**, *26*, 6063. https://doi.org/10.3390/molecules26196063.
8. Dabo, K.F.; Chèné, C.; Fameau, A.-L.; Karoui, R. Whipping Creams: Advances in Molecular Composition and Nutritional Chemistry. *Molecules* **2024**, *29*, 5933. https://doi.org/10.3390/molecules29245933.

References

1. Paggio, F.; Ritota, M.; Di Costanzo, M.G.; Barzaghi, S.; Monti, L.; Ulrici, A.; Manzi, P. Effects of time and temperature of storage on chemical and nutritional characteristics of raw milk for Provolone Valpadana PDO cheesemaking: A multivariate approach. *J. Dairy Res.* **2023**, *90*, 191–199. [CrossRef] [PubMed]
2. Ritota, M.; Di Costanzo, M.G.; Rampanti, G.; Aquilanti, L.; De León, C.M.B.; Tejada, L.; Psomas, A.; Dridi, B.A.M.; Manzi, P. Innovative thistle-curdled cheeses from the Mediterranean area: Nutritional evaluation of some relevant compounds. *J. Dairy Res.* **2025**, *92*, 107–116. [CrossRef] [PubMed]
3. International Dairy Federation. Dairy Matrix: The Case of Milk (Factsheet of the IDF N° 33/2023). 2023. Available online: https://shop.fil-idf.org/products/factsheet-of-the-idf-n-33-2023-dairy-matrix-the-case-of-milk (accessed on 20 February 2026).
4. Park, Y.W. *Bioactive Components in Milk and Dairy Products*, 1st ed.; John Wiley & Sons: Ames, IA, USA, 2009; pp. 1–426.
5. Pilla, F.; Valentini, A.; Lenstra, J.A.; Martin, P. Animal genetics and functional food. In *Livestock Farming Systems*, 1st ed.; Rubino, R., Sepe, L., Dimitriadou, A., Gibon, A., Eds.; Wageningen Academic Publishers: Wageningen, The Netherlands, 2006; pp. 49–59.
6. Morand-Fehr, P.; Fedele, V.; Decandia, M.; Le Frileux, Y. Influence of farming and feeding systems on composition and quality of goat and sheep milk. *Small Rumin. Res.* **2007**, *68*, 20–34. [CrossRef]
7. Delorme, M.M.; Guimarães, J.T.; Coutinho, N.M.; Balthazar, C.F.; Rocha, R.S.; Silva, R.; Margalho, L.P.; Pimentel, T.C.; Silva, M.C.; Freitas, M.Q.; et al. Ultraviolet radiation: An interesting technology to preserve quality and safety of milk and dairy foods. *Trends Food Sci. Technol.* **2020**, *102*, 146–154. [CrossRef]
8. Gemechu, T. Review on lactic acid bacteria function in milk fermentation and preservation. *Afr. J. Food Sci.* **2015**, *9*, 170–175. [CrossRef]
9. Ritota, M.; Mattera, M.; Di Costanzo, M.G.; Manzi, P. Evaluation of Crocins in Cheeses made with Saffron by UHPLC. *J. Braz. Chem. Soc.* **2018**, *29*, 248–257. [CrossRef]
10. Vélez, M.A.; Perotti, M.C.; Santiago, L.; Gennaro, A.M.; Hynes, E. Bioactive compounds delivery using nanotechnology: Design and applications in dairy food. In *Nutrient Delivery*, 1st ed.; Mihai Grumezescu, A., Ed.; Academic Press: Cambridge, MA, USA, 2017; pp. 221–250.
11. Ribeiro, N.G.; Xavier-Santos, D.; Campelo, P.H.; Guimarães, J.T.; Pimentel, T.C.; Duarte, M.C.K.; Freitas, M.Q.; Esmerino, E.A.; Silva, M.C.; Cruz, A.G. Dairy foods and novel thermal and non-thermal processing: A bibliometric analysis. *Innov. Food Sci. Emerg. Technol.* **2022**, *76*, 102934. [CrossRef]
12. Gómez-Cortés, P.; Juárez, M.; de la Fuente, M.A. Milk fatty acids and potential health benefits: An updated vision. *Trends Food Sci. Technol.* **2018**, *81*, 1–9. [CrossRef]

13. Samková, E.; Špička, J.; Hanuš, O.; Roubal, P.; Pecová, L.; Hasoňová, L.; Smetana, P.; Klimešová, M.; Čítek, J. Comparison of Fatty Acid Proportions Determined by Mid-Infrared Spectroscopy and Gas Chromatography in Bulk and Individual Milk Samples. *Animals* **2020**, *10*, 1095. [CrossRef] [PubMed]
14. De Kruif, C.G.; Huppertz, T.; Urban, V.S.; Petukhov, A.V. Casein micelles and their internal structure. *Adv. Colloid Interface Sci.* **2012**, *171–172*, 36–52. [CrossRef] [PubMed]
15. Sebastiani, C.; Arcangeli, C.; Ciullo, M.; Torricelli, M.; Cinti, G.; Fisichella, S.; Biagetti, M. Frequencies Evaluation of beta-Casein Gene Polymorphisms in Dairy Cows Reared in Central Italy. *Animals* **2020**, *10*, 252. [CrossRef] [PubMed]
16. Jianqin, S.; Leiming, X.; Lu, X.; Yelland, G.W.; Ni, J.; Clarke, A.J. Effects of milk containing only A2 beta casein versus milk containing both A1 and A2 beta casein proteins on gastrointestinal physiology, symptoms of discomfort, and cognitive behavior of people with self-reported intolerance to traditional cows' milk. *Nutr. J.* **2016**, *15*, 35. [CrossRef] [PubMed]
17. McLachlan, C.N.S. β-casein A1, ischaemic heart disease mortality, and other illnesses. *Med. Hypotheses* **2001**, *56*, 262–272. [CrossRef] [PubMed]
18. Coni, E.; Bocca, A.; Ianni, D.; Caroli, S. Preliminary evaluation of the factors influencing the trace element content of milk and dairy products. *Food Chem.* **1995**, *52*, 123–130. [CrossRef]
19. EU. Regulation (EU) No 1169/2011 of the European Parliament and of the Council of 25 October 2011 on the Provision of Food Information to Consumers. Available online: https://eur-lex.europa.eu/legal-content/EN/ALL/?uri=CELEX%3A32011R1169 (accessed on 20 February 2026).
20. EFSA, Panel on Dietetic Products, Nutrition and Allergies (NDA). Scientific opinion on the substantiation of health claims related to vitamin D and normal function of the immune system and inflammatory response (ID 154, 159), maintenance of normal muscle function (ID 155) and maintenance of normal cardiovascular function (ID 159) pursuant to Article 13(1) of Regulation (EC) No 1924/2006. *EFSA J.* **2010**, *8*, 1468. [CrossRef]
21. Institute of Medicine. *Dietary Reference Intakes for Calcium and Vitamin D*; National Academies Press: Washington, DC, USA, 2010.
22. EFSA, Panel on Dietetic Products, Nutrition and Allergies (NDA). Scientific opinion on the substantiation of a health claim related to vitamin D and contribution to the normal function of the immune system pursuant to Article 14 of Regulation (EC) No 1924/2006. *EFSA J.* **2015**, *13*, 4096. [CrossRef]
23. Scientific Advisory Committee on Nutrition (SACN). Vitamin D and Health Report Public Health England. 2016. Available online: https://www.gov.uk/government/publications/sacn-vitamin-d-and-health-report (accessed on 20 February 2026).
24. Schmid, A.; Walther, B. Natural Vitamin D Content in Animal Products. *Adv. Nutr.* **2013**, *4*, 453–462. [CrossRef] [PubMed]
25. Kasalová, E.; Aufartová, J.; Kujovská Krčmová, L.; Solichová, D.; Solich, P. Recent trends in the analysis of vitamin D and its metabolites in milk—A review. *Food Chem.* **2015**, *171*, 177–190. [CrossRef] [PubMed]
26. Itkonen, S.; Erkkola, M.; Lamberg-Allardt, C. Vitamin D Fortification of Fluid Milk Products and Their Contribution to Vitamin D Intake and Vitamin D Status in Observational Studies—A Review. *Nutrients* **2018**, *10*, 1054. [CrossRef] [PubMed]

https://doi.org/10.3390/molecules31060919

molecules

MDPI

Article

Predictions of Milk Fatty Acid Contents by Mid-Infrared Spectroscopy in Chinese Holstein Cows

Xiuxin Zhao [1,2,†], Yuetong Song [1,3,4,†], Yuanpei Zhang [2], Gaozhan Cai [2], Guanghui Xue [2], Yan Liu [2], Kewei Chen [4], Fan Zhang [1], Kun Wang [1,4], Miao Zhang [1,4], Yundong Gao [1], Dongxiao Sun [3,*], Xiao Wang [1,*] and Jianbin Li [1,*]

1 Institute of Animal Science and Veterinary Medicine, Shandong Academy of Agricultural Sciences, Jinan 250100, China
2 Shandong OX Livestock Breeding Co., Ltd., Jinan 250100, China
3 Department of Animal Genetics and Breeding, College of Animal Science and Technology, China Agricultural University, Beijing 100193, China
4 Yantai Institute, China Agricultural University, Yantai 264670, China
* Correspondence: sundx@cau.edu.cn (D.S.); xiaowangzntc@163.com (X.W.); msdljb@163.com (J.L.)
† These authors contributed equally to this work.

Abstract: Genetic improvement of milk fatty acid content traits in dairy cattle is of great significance. However, chromatography-based methods to measure milk fatty acid content have several disadvantages. Thus, quick and accurate predictions of various milk fatty acid contents based on the mid-infrared spectrum (MIRS) from dairy herd improvement (DHI) data are essential and meaningful to expand the amount of phenotypic data available. In this study, 24 kinds of milk fatty acid concentrations were measured from the milk samples of 336 Holstein cows in Shandong Province, China, using the gas chromatography (GC) technique, which simultaneously produced MIRS values for the prediction of fatty acids. After quantification by the GC technique, milk fatty acid contents expressed as g/100 g of milk (milk-basis) and g/100 g of fat (fat-basis) were processed by five spectral pre-processing algorithms: first-order derivative (DER1), second-order derivative (DER2), multiple scattering correction (MSC), standard normal transform (SNV), and Savitzky–Golsy convolution smoothing (SG), and four regression models: random forest regression (RFR), partial least square regression (PLSR), least absolute shrinkage and selection operator regression (LassoR), and ridge regression (RidgeR). Two ranges of wavebands (4000~400 cm^{-1} and 3017~2823 cm^{-1}/1805~1734 cm^{-1}) were also used in the above analysis. The prediction accuracy was evaluated using a 10-fold cross validation procedure, with the ratio of the training set and the test set as 3:1, where the determination coefficient (R^2) and residual predictive deviation (RPD) were used for evaluations. The results showed that 17 out of 31 milk fatty acids were accurately predicted using MIRS, with RPD values higher than 2 and R^2 values higher than 0.75. In addition, 16 out of 31 fatty acids were accurately predicted by RFR, indicating that the ensemble learning model potentially resulted in a higher prediction accuracy. Meanwhile, DER1, DER2 and SG pre-processing algorithms led to high prediction accuracy for most fatty acids. In summary, these results imply that the application of MIRS to predict the fatty acid contents of milk is feasible.

Keywords: prediction; milk; fatty acid content; mid-infrared spectroscopy; Chinese Holstein cow

1. Introduction

Lipids in milk provide a major source of energy and the essential structural components for the cell membranes of the newborns in all mammalian species. They also confer distinctive properties to dairy foods that affect further processing procedures [1]. Milk fat is rich in many fatty acids that are important to human health [2–4]. Studies have shown that more than 400 different fatty acids have been identified in milk fat, but most of them only appeared in trace amounts [5], where around 12 kinds of fatty acids in bovine milk fat presented at above a 1% concentration [6]. Moreover, changes in milk fatty acids may also affect cow health and energy statuses [7].

Currently, several techniques have been developed to measure fatty acids in milk, including high performance liquid chromatography (HPLC), gas chromatography (GC), near-infrared spectroscopy (NIRS), mid-infrared spectrum (MIRS), etc. [8–10]. Chemical methods (e.g., HPLC and GC) provide high measurement accuracy for fatty acid contents of bovine milk, but their pretreatments are multifarious and costly, causing difficulties in realizing the high-throughput measurements [11,12]. Of note, infrared spectroscopy-based measurement methods show advantages of providing rapid and low-cost predictions of milk fatty acid contents [13]; thus, they have become the promising technologies for high-throughput measurements, but they still need to be optimized to improve their prediction accuracy.

The utilization of infrared spectroscopy to predict the milk fatty acid contents in dairy cattle has been reported in many studies. Coppa et al. (2010) established a prediction equation for milk fatty acid contents based on the NIRS from 468 milk samples that predicted the total milk fatty acids, SFA, MUFA, PUFA, C18:1, and conjugated linoleic acid (CLA), with R^2 values greater than 0.88. Soyeurt et al. (2006) developed a fatty acid prediction model using 600 milk samples from 275 cows of 6 breeds to predict C10:0, C12:0, C14:0, C16:0, C16:1cis-9, C18:1, C18:2cis-9, SFA (saturated fatty acids), and MUFA (monounsaturated fatty acids), based on MIRS data, with the cross-validated coefficients of determination (R^2) of 0.62 ~ 0.94. Subsequently, Soyeurt et al. (2011) investigated the MIRS prediction of fatty acids across various cattle breeds, production systems, and countries. They summarized that the usefulness of the built equations providing the best prediction accuracy for animal breeding and milk payment systems was $R^2 \geq 0.75$ and 0.95, respectively [4]. For the Dutch cattle breeds (Dutch Friesian, Meuse-Rhine-Yssel, Groningen White Headed, and Jersey), Maurice-Van Eijndhoven et al. (2020) updated the calibration equations from the European project RobustMilk [4] using the enlarged datasets and validated their usefulness to predict most milk fatty acids. De Marchi et al. (2011) used 267 milk samples from Brown Swiss cattle to predict fatty acids by MIRS and suggested the implementation of the used prediction models in milk recording schemes on fatty acid contents information for breeding purposes. Fleming et al. (2017) used MIRS to predict fatty acid contents from 373 cows of four breeds and obtained the cross-validation R^2 of 0.60~0.90 for most individual fatty acid models. In addition, the genetic correlations among milk fatty acids predicted by MIRS were also explored in a large-scale milk sampling (n = 34,141) of New Zealand dairy cattle, where they implied the application of MIRS as the phenotypic proxy for the genetic selection of fatty acid contents [14]. In the Chinese Holstein population, Du et al. (2020) estimated the heritability of MIRS and several milk production traits, i.e., protein, fat, and lactose percentages, along with their genetic correlations. They found that MIRS heritability ranged from 0 to 0.11 and genetic correlations varied significantly [15]. In sheep, ewes, and goats, MIRS was also used to predict the fatty acid profiles for the establishment and validation of the predictive models [16–18].

Previous studies used a partial least square regression model (non-integrated learning model) [19,20] to investigate the effects of different spectral preprocessing methods on the prediction equation accuracy [4,5,21–23]. However, the combined effects of the regression models and spectral preprocessing methods on the prediction equation accuracy for different fatty acids has rarely been explored, especially for the milk fat of Chinese Holstein cows. Therefore, the objective of this study was to investigate the prediction methods under the optimal strategy to predict milk fatty acids with high accuracy based on the MIRS data from the dairy herd improvement (DHI) database of Chinese Holstein cattle and to potentially provide the high-throughput measurements of a large amount of milk fatty acid phenotypic data; thereby, our study enabled milk fatty acid traits to be feasibly recorded for genetic evaluations of such traits in dairy cattle breeding programs in China. To the best of our knowledge, this is the first time the MIRS predictions on fatty acids of two types of fatty acid measurements (g/100 g of milk and g/100 g fat) have been investigated with five pre-processed algorithms and two ranges of wavebands (4000~400 cm^{-1} and 3017~2823 cm^{-1}/1805~1734 cm^{-1}) using four regression models in Chinese Holstein cattle.

2. Results

2.1. Statistical Description

After quantification by the GC technique, statistical descriptions of individual and grouped fatty acid contents expressed as milk-basis (g/100 g of milk) and fat-basis (g/100 g of fat) are summarized in Table 1. The mean values of the individual fatty acid contents varied from 0.003 (C11:0, C20:1, C20:5n3, and C18:3n6) to 0.877 (C16:0) and their variation coefficients varied from 5.837% (C24:0) to 35.416% (C10:0), when they were expressed as milk-basis (g/100 g of milk). For grouped fatty acid contents, the mean values varied from 0.060 (SCFA) to 1.627 (SFA), and their variation coefficients ranged from 25.514% (PUFA) to 33.392% (SCFA) (Table 1). Similarly, the mean values of individual fatty acid contents varied from 0.094 (C20:5n3) to 28.620 (C16:0), and their variation coefficients varied from 13.802% (C16:0) to 44.207% (C22:1n9), when they were expressed as fat-basis (g/100 g of fat). For grouped fatty acid contents, the mean values varied from 1.934 (SCFA) to 52.710 (SFA), and their variation coefficients ranged from 12.978% (MCFA) to 19.365% (LCFA) (Table 1).

Table 1. The minimum, mean, maximum, and variation coefficient of fatty acid contents measured by the GC technique.

Fatty Acid	Milk-Basis (g/100 g of Milk)					Fat-Basis (g/100 g of Fat)				
	Sample Size	Minimum	Mean	Maximum	Variation Coefficient (%)	Sample Size	Minimum	Mean	Maximum	Variation Coefficient (%)
C8:0	325	0.007	0.016	0.028	28.784	324	0.327	0.532	0.757	14.778
C10:0	323	0.013	0.044	0.082	35.416	326	0.598	1.402	2.324	21.273
C11:0	317	0.002	0.003	0.006	25.439	319	0.053	0.110	0.175	23.125
C12:0	321	0.019	0.062	0.115	34.871	327	0.829	2.018	3.323	21.721
C13:0	321	0.003	0.005	0.008	24.468	323	0.082	0.166	0.255	22.062
C14:0	322	0.094	0.231	0.371	27.975	321	4.058	7.546	11.358	15.075
C15:0	320	0.012	0.029	0.052	29.138	321	0.456	0.968	1.519	21.227
C16:0	325	0.366	0.877	1.491	28.154	323	17.710	28.620	42.251	13.802
C17:0	324	0.008	0.016	0.027	26.180	322	0.285	0.528	0.797	19.289
C18:0	320	0.098	0.313	0.600	35.024	326	4.070	10.254	17.150	25.195
C20:0	321	0.006	0.008	0.011	15.098	321	0.157	0.270	0.398	19.278
C22:0	332	0.004	0.005	0.006	9.413	329	0.074	0.172	0.286	25.529
C24:0	324	0.004	0.005	0.005	5.837	323	0.069	0.154	0.246	24.812
C14:1	325	0.006	0.019	0.033	32.157	317	0.230	0.610	1.138	31.033
C16:1	322	0.016	0.038	0.068	30.326	320	0.602	1.258	2.202	26.094
C18:1n9c	321	0.189	0.460	0.815	28.657	323	7.953	15.282	23.746	20.201
C20:1	321	0.003	0.003	0.005	13.134	319	0.065	0.116	0.190	24.333
C22:1n9	322	0.007	0.015	0.028	32.090	317	0.193	0.510	1.140	44.207
C20:3n6	322	0.003	0.006	0.009	24.212	322	0.109	0.192	0.288	18.233
C20:4n6	323	0.004	0.007	0.010	20.368	317	0.122	0.222	0.329	18.416
C20:5n3	332	0.002	0.003	0.004	10.163	323	0.046	0.094	0.149	23.530
C18:2n6c	323	0.030	0.070	0.120	28.437	327	1.052	2.300	3.654	18.149
C18:3n6	321	0.003	0.003	0.004	7.621	318	0.047	0.100	0.169	24.898
C18:3n3	324	0.004	0.008	0.013	24.352	324	0.155	0.278	0.417	16.480
SFA	325	0.692	1.627	2.714	28.415	322	29.494	52.710	74.351	13.287
UFA	323	0.288	0.638	1.090	26.277	326	13.162	21.266	31.904	18.082
MUFA	322	0.240	0.539	0.938	26.878	324	9.501	17.920	27.111	19.321
PUFA	324	0.046	0.098	0.159	25.514	325	1.720	3.196	4.863	16.229
SCFA	323	0.020	0.060	0.109	33.392	324	0.926	1.934	2.936	18.762
MCFA	325	0.580	1.269	2.147	27.475	322	25.617	41.372	61.272	12.978
LCFA	321	0.371	0.925	1.610	28.353	326	17.003	30.776	46.784	19.365

Note: SFA, UFA, MUFA, PUFA, SCFA, MCFA, LCFA, and GC indicate saturated fatty acid, unsaturated fatty acid, monounsaturated fatty acid, polyunsaturated fatty acid, short chain (4 to 10 carbons) fatty acid, medium chain (11 to 16 carbons) fatty acid, long chain (more than 16 carbons) fatty acid, and gas chromatography, respectively. The variation coefficient (%) is the ratio of standard deviation to the mean, which can be used to compare the degree of dispersion among the fatty acids.

2.2. Predictions of Milk Fatty Acid Contents

The best prediction accuracy obtained by the optimal strategy from the test set for each fatty acid is summarized in Table 2, after considering different pre-processing algorithms, MIRS ranges, and regression models. In total, 16, 7, 6, and 2 fatty acids achieved the best prediction accuracy when the RFR, LassoR, PLSR, and RidgeR models were used, respectively. Similarly, the DER2, DER1, SG, SNV, and MSC algorithms resulted in 9, 8, 8, 4, and 2 fatty acids for best prediction accuracy, respectively. In addition, 22 fatty acids obtained the best prediction accuracy when they were expressed as g/100 g of milk (milk-basis), but only 9 fatty acids when expressed as g/100 g of fat (fat-basis). For most fatty acids (16/31), the ensemble learning model (RFR), with higher robustness and generalization, produced higher prediction accuracy than those predicted by the non-ensemble learning models (Table 2).

Table 2. Best prediction accuracy for the optimal strategy in the test set for each fatty acid.

Fatty Acid	Pre-Processing Algorithm	MIRS Range (cm^{-1})	Model	Basis (g/100 g)	Test Set	
					R^2	RPD
C8:0	SNV	3017~2823/1805~1734	PLSR	Milk	0.77	2.11
C10:0	DER1	3017~2823/1805~1734	RFR	Milk	0.77	2.07
C11:0	DER1	3017~2823/1805~1734	LassoR	Fat	0.55	1.48
C12:0	DER1	3017~2823/1805~1734	LassoR	Milk	0.84	2.50
C13:0	SG	3017~2823/1805~1734	PLSR	Milk	0.66	1.72
C14:0	DER1	4000~400	RFR	Milk	0.78	2.05
C15:0	SG	3017~2823/1805~1734	PLSR	Milk	0.57	1.53
C16:0	SG	3017~2823/1805~1734	RFR	Milk	0.75	1.98
C17:0	SG	3017~2823/1805~1734	LassoR	Milk	0.73	1.89
C18:0	DER1	4000~400	PLSR	Milk	0.77	2.08
C20:0	SNV	3017~2823/1805~1734	PLSR	Fat	0.82	2.35
C22:0	DER2	4000~400	RFR	Fat	0.86	2.66
C24:0	SG	4000~400	RFR	Fat	0.80	2.20
C14:1	MSC	3017~2823/1805~1734	PLSR	Fat	0.62	1.63
C16:1	SNV	3017~2823/1805~1734	LassoR	Milk	0.62	1.64
C18:1n9c	SG	3017~2823/1805~1734	LassoR	Milk	0.77	2.00
C20:1	DER2	4000~400	RFR	Fat	0.76	2.04
C22:1n9	DER1	4000~400	RFR	Fat	0.65	1.67
C18:2n6c	MSC	4000~400	RFR	Milk	0.63	1.61
C18:3n3	SG	4000~400	RFR	Milk	0.70	1.82
C18:3n6	DER2	3017~2823/1805~1734	RFR	Fat	0.76	2.00
C20:3n6	DER1	4000~400	RFR	Milk	0.62	1.61
C20:4n6	SNV	4000~400	RFR	Milk	0.50	1.42
C20:5n3	DER1	4000~400	RFR	Fat	0.91	3.06
SFA	SG	3017~2823/1805~1734	RFR	Milk	0.76	2.01
UFA	DER2	3017~2823/1805~1734	LassoR	Milk	0.82	2.15
MUFA	DER2	3017~2823/1805~1734	LassoR	Milk	0.79	2.06
PUFA	DER2	4000~400	RidgeR	Milk	0.71	1.75
SCFA	DER2	4000~400	RFR	Milk	0.77	2.04
MCFA	DER2	3017~2823/1805~1734	RFR	Milk	0.75	2.00
LCFA	DER2	3017~2823/1805~1734	RidgeR	Milk	0.83	2.29

Note: MIRS, DER1, DER2, MSC, SNV, SG, RFR, PLSR, LassoR, RidgeR, R^2, and RPD indicate mid-infrared spectrum, first-order derivative, second-order derivative, multiple scattering correction, standard normal transform, Savitzky–Golsy convolution smoothing, random forest regression, partial least square regression, least absolute shrinkage and selection operator regression, ridge regression, determination coefficient, and residual predictive deviation, respectively.

In Table 2, the best prediction accuracy (R^2) for the optimal strategy showed R^2 values from 0.62 (C20.3n6) to 0.91 (C20.5n3) in the test set for 28 fatty acids, where only 6 fatty acids showed R^2 values higher than 0.8, including C12:0 (0.84), C20:0 (0.82), C22:0 (0.86), C20:5n3 (0.91), UFA (0.82), and LCFA (0.83). For R^2 values higher than 0.75 and RPD values higher than 2, we found 17 fatty acids in total: C8:0 (R^2 = 0.77 and RPD = 2.11),

C10:0 (R^2 = 0.77 and RPD = 2.07), C12:0 (R^2 = 0.84 and RPD = 2.50), C14:0 (R^2 = 0.78 and RPD = 2.05), C18:0 (R^2 = 0.77 and RPD = 2.08), C20:0 (R^2 = 0.82 and RPD = 2.35), C22:0 (R^2 = 0.86 and RPD = 2.66), C24:0 (R^2 = 0.80 and RPD = 2.20), C18:1n9c (R^2 = 0.77 and RPD = 2.00), C20:1 (R^2 = 0.76 and RPD = 2.04), C20:5n3 (R^2 = 0.91 and RPD = 3.06), SFA (R^2 = 0.76 and RPD = 2.01), UFA (R^2 = 0.82 and RPD = 2.15), MUFA (R^2 = 0.79 and RPD = 2.06), SCFA (R^2 = 0.77 and RPD = 2.04), MCFA (R^2 = 0.75 and RPD = 2.00), and LCFA (R^2 = 0.83 and RPD = 2.29) (Table 2).

Table 3 shows the best prediction accuracy of different prediction models for each fatty acid, using training and test sets. All prediction accuracies (R^2 and RPD) after four regression model analyses (RFR, PLSR, LassoR, and RidgeR), based on two types of fatty acid measurements (g/100 g of milk and g/100 g fat), two ranges of wavebands (4000~400 cm^{-1} and 3017~2823 cm^{-1}/1805~1734 cm^{-1}), and five spectral pre-processing algorithms (DER1, DER2, MSC, SNV, and SG), are listed in Supplementary File S1. In the training set, the R^2 values ranged from 0.18 to 0.79, with a mean of 0.58, and RPD values ranging from 1.08 to 2.18, with a mean of 1.59, when expressed as milk-basis (g/100 g of milk). Similarly, R^2 values ranged from 0.13 to 0.90, with a mean of 0.47, and RPD values from 1.07 to 3.20, with mean of 1.52, when expressed as fat-basis (g/100 g of fat). In the test set, R^2 values ranged from 0.14 to 0.84 with mean of 0.66 and RPD values from 1.04 to 2.50 with mean of 1.78 when expressed as milk basis (g/100 g of milk). Similarly, the R^2 values ranged from 0.15 to 0.91 with mean of 0.49 and RPD values from 1.07 to 3.06 with mean of 1.52 when expressed as fat basis (g/100 g of fat) (Table 3). Additionally, the MIRS and processed MIRS after DER1, DER2, and SG pre-processing algorithms are shown in Figure 1.

Table 3. Best prediction accuracy of different prediction models for each fatty acid expressed as g/100 g of fat and g/100 g of milk, using training and test sets.

Fatty Acid	Pre-Processing Algorithm		MIRS Range (cm^{-1})		Model		Training Set				Test Set			
							R^2		RPD		R^2		RPD	
	Milk	Fat	Milk	Fat	Milk	Fat	Milk	Fat	Milk	Fat	Milk	Fat	Milk	Fat
C8:0	SNV	MSC	3017~2823/1805~1734	3017~2823/1805~1734	PLSR	LassoR	0.75	0.43	2.01	1.33	0.77	0.43	2.11	1.32
C10:0	DER1	DER1	3017~2823/1805~1734	3017~2823/1805~1734	RFR	LassoR	0.61	0.49	1.60	1.40	0.77	0.44	2.07	1.33
C11:0	DER2	DER1	3017~2823/1805~1734	3017~2823/1805~1734	LassoR	LassoR	0.57	0.51	1.53	1.43	0.53	0.55	1.46	1.48
C12:0	DER1	SNV	3017~2823/1805~1734	3017~2823/1805~1734	LassoR	LassoR	0.79	0.55	2.18	1.49	0.84	0.27	2.50	1.17
C13:0	SG	SNV	3017~2823/1805~1734	3017~2823/1805~1734	PLSR	LassoR	0.24	0.56	1.16	1.50	0.66	0.42	1.72	1.30
C14:0	DER1	DER1	4000~400	3017~2823/1805~1734	RFR	PLSR	0.66	0.16	1.72	1.10	0.78	0.43	2.05	1.34
C15:0	SG	MSC	3017~2823/1805~1734	3017~2823/1805~1734	PLSR	PLSR	0.45	0.25	1.37	1.17	0.57	0.32	1.53	1.22
C16:0	SG	DER2	3017~2823/1805~1734	4000~400	RFR	RidgeR	0.64	0.55	1.66	1.33	0.75	0.22	1.98	1.12
C17:0	SG	MSC	3017~2823/1805~1734	3017~2823/1805~1734	LassoR	PLSR	0.65	0.40	1.70	1.32	0.73	0.59	1.89	1.56
C18:0	DER1	SNV	4000~400	3017~2823/1805~1734	PLSR	LassoR	0.66	0.60	1.72	1.58	0.77	0.55	2.08	1.49
C20:0	SG	SNV	3017~2823/1805~1734	3017~2823/1805~1734	PLSR	PLSR	0.52	0.76	1.46	2.04	0.71	0.82	1.88	2.35
C22:0	DER2	DER2	4000~400	4000~400	RidgeR	RFR	0.70	0.83	1.76	2.42	0.52	0.86	1.44	2.66
C24:0	DER2	SG	4000~400	4000~400	RidgeR	RFR	0.64	0.90	1.55	3.20	0.61	0.80	1.46	2.20
C14:1	SNV	MSC	3017~2823/1805~1734	3017~2823/1805~1734	LassoR	PLSR	0.63	0.38	1.65	1.28	0.51	0.62	1.40	1.63
C16:1	SNV	MSC	3017~2823/1805~1734	3017~2823/1805~1734	LassoR	LassoR	0.54	0.38	1.47	1.27	0.62	0.55	1.64	1.50
C18:1n9c	SG	MSC	3017~2823/1805~1734	3017~2823/1805~1734	LassoR	LassoR	0.60	0.52	1.58	1.45	0.77	0.34	2.00	1.20
C20:1	SG	DER2	3017~2823/1805~1734	4000~400	PLSR	RFR	0.54	0.77	1.48	2.06	0.49	0.76	1.41	2.04
C22:1n9	DER2	DER1	4000~400	4000~400	RFR	RFR	0.51	0.53	1.43	1.45	0.45	0.65	1.36	1.67
C18:2n6c	MSC	SG	4000~400	4000~400	RFR	RidgeR	0.59	0.13	1.56	1.07	0.63	0.15	1.61	1.08
C18:3n3	SG	DER1	4000~400	3017~2823/1805~1734	RFR	RFR	0.60	0.17	1.59	1.09	0.70	0.27	1.82	1.13
C18:3n6	SG	DER2	3017~2823/1805~1734	3017~2823/1805~1734	RFR	RFR	0.18	0.84	1.08	2.47	0.14	0.76	1.04	2.00
C20:3n6	DER1	MSC	4000~400	3017~2823/1805~1734	RFR	PLSR	0.50	0.23	1.42	1.15	0.62	0.39	1.61	1.29
C20:4n6	SNV	SNV	4000~400	4000~400	RFR	PLSR	0.44	0.29	1.34	1.19	0.50	0.46	1.42	1.37
C20:5n3	DER2	DER1	4000~400	4000~400	RFR	RFR	0.33	0.83	1.23	2.41	0.43	0.91	1.29	3.06
LCFA	DER2	DER1	3017~2823/1805~1734	4000~400	RidgeR	RFR	0.68	0.41	1.78	1.31	0.83	0.42	2.29	1.32
MCFA	DER2	SNV	3017~2823/1805~1734	3017~2823/1805~1734	RFR	LassoR	0.64	0.23	1.67	1.14	0.75	0.28	2.00	1.18
MUFA	DER2	DER1	3017~2823/1805~1734	3017~2823/1805~1734	LassoR	LassoR	0.61	0.56	1.59	1.51	0.79	0.43	2.06	1.30
PUFA	DER2	SG	4000~400	3017~2823/1805~1734	RidgeR	RFR	0.71	0.16	1.80	1.08	0.71	0.16	1.75	1.07
SCFA	DER2	MSC	4000~400	3017~2823/1805~1734	RFR	LassoR	0.66	0.51	1.71	1.43	0.77	0.48	2.04	1.37
SFA	SG	SG	3017~2823/1805~1734	3017~2823/1805~1734	RFR	LassoR	0.66	0.32	1.73	1.21	0.76	0.25	2.01	1.16
UFA	DER2	MSC	3017~2823/1805~1734	3017~2823/1805~1734	LassoR	LassoR	0.62	0.42	1.62	1.31	0.82	0.48	2.15	1.38

Note: MIRS, DER1, DER2, MSC, SNV, SG, RFR, PLSR, LassoR, RidgeR, R^2, and RPD indicate mid-infrared spectrum, first-order derivative, second-order derivative, multiple scattering correction, standard normal transform, Savitzky–Golsy convolution smoothing, random forest regression, partial least square regression, least absolute shrinkage and selection operator regression, ridge regression, determination coefficient, and residual predictive deviation, respectively.

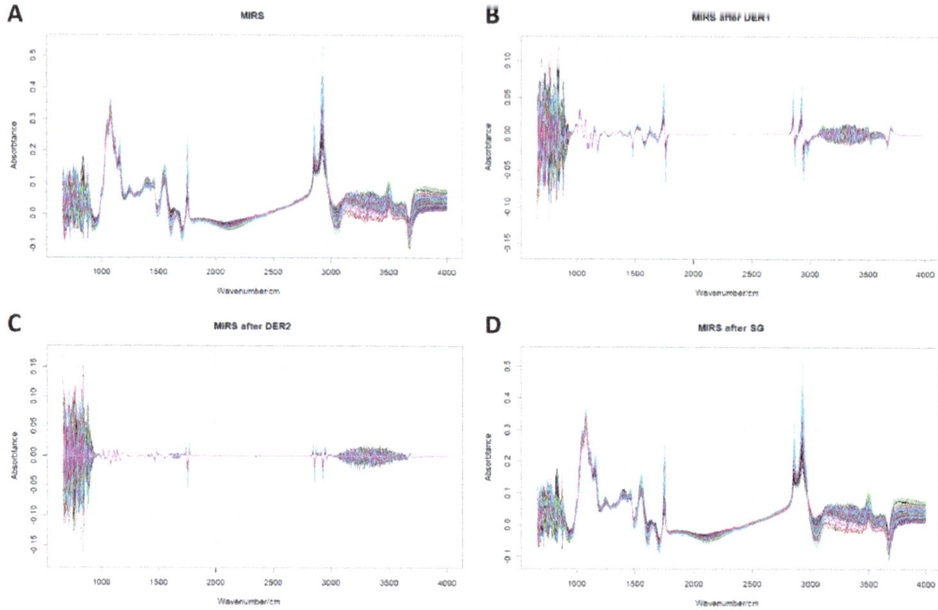

Figure 1. MIRS after DER1, DER2, and SG pre-processing algorithms. Note: MIRS, DER1, DER2, and SG indicate mid-infrared spectrum, first-order derivative, second-order derivative, and Savitzky–Golsy convolution smoothing, respectively.

3. Materials and Methods

3.1. Milk Samples and Fatty Acids

Milk samples were collected from 336 Holstein cows on a farm in Shandong Province, China, including one small tube (30 mL) and one large tube (50 mL) from each cow. After sampling, all tubes were immediately stored in liquid nitrogen (−196 °C) and delivered to our experimental lab for further analysis within 6 h. In this study, to maintain analysis consistency, none of the 672 collected milk samples received any preservative additions, and the milk samples in the 30 mL and 50 mL tubes were used to measure fatty acid contents and MIRS, respectively.

A total of 24 kinds of fatty acid contents, which included C8:0, C10:0, C11:0, C12:0, C13:0, C14:0, C14:1, C15:0, C16:0, C16:1, C17:0, C18:0, C18:1n9c, C18:2n6c, C20:0, C18:3n6, C18:3n3, C20:1, C22:0, C20:3n6, C20:4n6, C22:1n9, C20:5n3, and C24:0, were measured and quantified in each milk sample from the 30 mL tubes using the GC technique. Due to the limitations of GC technique, the apparent missing values were replaced by the averaged values of the whole fatty acids that had been quantified by the GC technique. The outliers of quality control for the fatty acids were defined by the mean reference values ± two standard deviations. For each milk sample from the 50 mL tubes, 899 raw data points for MIRS values in the complete waveband range of 4000~400 cm^{-1} were obtained by Bentley spectrometers (Bentley Instruments Inc., Chaska, MN, United States), following the routine methodology (e.g., 30 min preheating and sufficient shaking before operation). Afterwards, additional raw MIRS values, as the measurement replicates, were also obtained using the same milk samples. Finally, two raw MIRS values were transformed by the Fourier method [24] for further pre-processing steps.

Here, the GC methodology for the quantification of fatty acid contents in our study was similar to those in other studies [4,25]. The outputs of the GC technique were generated by analyzing the methyl esters from the fat in the milk following ISO Standard 15884 (ISO–IDF (International Organization for Standardization–International Dairy Federation),

2002). Normally, the GC technique is used as the gold standard for fatty acid measurements because of its high accuracy, even for low contents [26,27], while the MIRS method is more rapid and less expensive [13,21].

According to the saturation conditions of hydrocarbon chains, fatty acids are classified as saturated fatty acids (SFAs), unsaturated fatty acids (UFAs), monounsaturated fatty acids (MUFAs), and polyunsaturated fatty acids (PUFAs) [5]. According to the carbon chain lengths, fatty acids are classified as short chain (4 to 10 carbons) fatty acids (SCFAs), medium chain (11 to 16 carbons) fatty acids (MCFAs), and long chain (more than 16 carbons) fatty acids (LCFAs). Consequently, 7 fatty acid groups for 24 kinds of the above fatty acids were obtained (Table 4).

Table 4. Seven classified fatty acid groups, according to hydrocarbon chain saturation and carbon chain length.

Fatty Acid Group According to Hydrocarbon Chain Saturation		Fatty Acid Group According to Carbon Chain Length	
SFA	C8:0, C10:0, C11:0, C12:0, C13:0, C14:0, C15:0, C16:0, C17:0, C18:0, C20:0, C22:0, C24:0	SCFA	C8:0, C10:0
UFA	C14:1, C16:1, C18:1n9c, C18:2n6c, C18:3n6, C18:3n3, C20:1, C20:3n6, C20:4n6, C22:1n9, C20:5n3	MCFA	C11:0, C12:0, C13:0, C14:0, C15:0, C16:0, C16:1
MUFA	C14:1, C16:1, C18:1n9c, C20:1, C22:1n9	LCFA	C17:0, C18:0, C18:1n9c, C18:2n6c, C20:0, C18:3n6, C18:3n3, C20:1, C22:0, C20:3n6, C20:4n6, C22:1n9, C20:5n3, C24:0
PUFA	C18:2n6t, C18:2n6c, C18:3n6, C18:3n3, C20:3n6, C20:4n6, C22:2, C20:5n3		

Note: SFA, UFA, MUFA, PUFA, SCFA, MCFA, and LCFA indicate saturated fatty acid, unsaturated fatty acid, monounsaturated fatty acid, polyunsaturated fatty acid, short chain (4 to 10 carbons) fatty acid, medium chain (11 to 16 carbons) fatty acid, and long chain (more than 16 carbons) fatty acid, respectively.

3.2. Predictions of Milk Fatty Acid Contents Using MIRS Data

Each fatty acid content quantified by the GC technique was converted from g/100 g of milk (milk-basis) to g/100 g of fat (fat-basis) using the fat contents determined by MIRS. The final MIRS values (the averaged values of two transformed MIRS replicates using the same milk sample) were processed using five spectral pre-processing algorithms, i.e., first-order derivative (DER1), second-order derivative (DER2), multiple scattering correction (MSC), standard normal transform (SNV), and Savitzky–Golsy convolution smoothing (SG). In order to compare the influence of each pre-processing algorithm, we used them individually to process the final MIRS values. Two types of fatty acid measurements (g/100 g of milk and g/100 g fat), with the five pre-processed spectra above and two ranges of wavebands (4000~400 cm^{-1} and 3017~2823 cm^{-1}/1805~1734 cm^{-1}), were analyzed using four regression models, i.e., random forest regression (RFR), partial least square regression (PLSR), least absolute shrinkage and selection operator regression (LassoR), and ridge regression (RidgeR). The determination coefficient (R^2) and residual predictive deviation (RPD) were used to evaluate the metrics of the four regression models. Prediction accuracy was assessed using a 10-fold cross validation procedure with the ratio of the training set and the test set as 3:1. The GC quantification technique, fatty acid measurements, spectral pre-processing algorithms, fatty acid prediction methods, and prediction accuracy assessments are summarized in Figure 2.

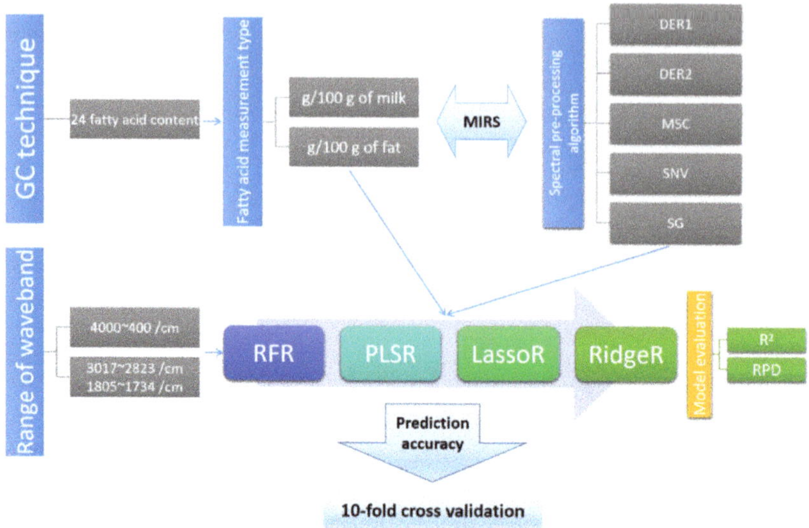

Figure 2. Summary of fatty acid prediction methods. Note: GC, MIRS, DER1, DER2, MSC, SNV, SG, RFR, PLSR, LassoR, RidgeR, R^2, and RPD indicate gas chromatography, mid-infrared spectrum, first-order derivative, second-order derivative, multiple scattering correction, standard normal transform, Savitzky–Golsy convolution smoothing, random forest regression, partial least square regression, least absolute shrinkage and selection operator regression, ridge regression, determination coefficient, and residual predictive deviation, respectively.

4. Discussion

The concentrations of different milk fatty acids in our study (Table 1) seem slightly lower than those in other studies [5,28–30], which could be caused mainly by the differences in feed diet and milk-collection times of the farm, where they supplied their own total mix ration (TMR) three times per day, which is less than other similar Chinese Holstein cattle farms (four or five times per day). Compared to the results of Soyeurt et al. (2011) and Fleming et al. (2017), the variation coefficients ranged from 12.978% to 44.207% as fat-basis (g/100 g of fat), which were slightly lower, on average, than those in other studies. The higher variations of fatty acids as fat-basis (g/100 g of fat) in relation to those as milk-basis (g/100 g of milk) could be a tendency in which fatty acids exhibited high mean values and variation coefficients (Table 1).

Many previous studies have investigated the accuracy and applicability of prediction models based on R^2 values. Soyeurt et al. (2011) suggested that models with $R^2 > 0.75$ might be utilized for animal breeding. However, Zaalberg et al. (2021) used prediction models with $R^2 > 0.6$ for mineral elements in animal breeding [31]. Cecchinato et al. (2009) showed low R^2 values for curd characteristics predicted by MIRS, but they found high genetic correlations between the measured values and the predicted values [32]. In our study, 17 fatty acids (C8:0, C10:0, C12:0, C14:0, C18:0, C20:0, C22:0, C24:0, C18:1n9c, C20:1, C20:5n3, SFA, UFA, MUFA, SCFA, MCFA, and LCFA) showed RPD ≥ 2 and $R^2 \geq 0.75$ (Table 2), which is consistent with the results of Soyeurt et al. (2006). This suggests that these 17 fatty acids can be accurately predicted using MIRS, and that this method has the potential for further fat trait selections in animal breeding. Furthermore, 6 fatty acids (C12:0, C20:0, C22:0, C20:5n3, UFA, and LCFA) with $R^2 > 0.8$, which were well predicted by MIRS, could also be used for breeding selections. For the grouped fatty acids, the R^2 values of the test set were greater than 0.7 (Table 3), which is consistent with the results of Soyeurt et al. (2006), Rutten et al. (2009), and Fleming et al. (2017). For both the training and the test sets, 6 individual fatty acids (C20:0, C22:0, C24:0, C20:1, C18:3n6, and C20:5n3) as fat-basis (g/100 g of fat) showed R^2 values greater than 0.7, whereas inconsistent results were found in other studies [4,5]. Fleming et al. (2017) obtained higher accuracy (R^2) from fatty acids expressed on the milk-basis than on the fat-basis. Soyeurt et al. (2011) used the fatty acids predicted in milk for their prediction in fat and only achieved results better than those of the direct prediction in fat for C6:0, C12:0, C18:2 cis-9, cis-12, SFA, and SCFA. RPD is also used to measure the prediction effect and accuracy of models [33,34]. Three classifications of RPD are as follows: high prediction accuracy, which can be used for the quantitative prediction of substances when RPD $\geq$ 2; good prediction, which can be used for rough quantitative prediction or qualitative analysis when 1.4 $\leq$ RPD < 2; and low prediction accuracy, which cannot be used for quantitative prediction when RPD < 1.4. Generally, a higher accuracy (R^2 and RPD) can also be observed in the prediction of fatty acids by MIRS on the milk-basis (n = 22) than on the fat-basis (n = 9) (Tables 2 and 3), which is consistent with the results of other studies [4,5,21,35].

Different spectral pre-processing algorithms influence the prediction accuracy of fatty acids. Soyeurt et al. (2012) used MIRS to predict the lactoferrin content in bovine milk and obtained the highest prediction accuracy using PLSR based on DER1. Our study also found that derivatives (DER1 and DER2) and SG smoothing algorithms can be applied for most fatty acid predictions (Table 2). The derivative algorithm uses the absorbance values corresponding to each of two adjacent wave points to calculate their derivative values, where the spectrum is processed by the derivative. The wave points with large differences in absorbance reduce signal/noise interference; then, the corresponding value of the current wave point moves sequentially to retain the spectral information for stronger spectrum continuity (Figure 1).

5. Conclusions

In this study, different regression models led to varying prediction accuracy of fatty acid contents, while different pre-processing algorithms for the spectra also influenced prediction accuracy. It was revealed that a higher accuracy for most fatty acids can be achieved when derivative and SG pre-processing algorithms for RFR models were used. Therefore, after a series of evaluations in Chinese Holstein cows, these results suggest that the application of MIRS to predict the fatty acid contents of milk is feasible.

Supplementary Materials: The following supporting information can be downloaded at: https://www.mdpi.com/article/10.3390/molecules28020666/s1, Supplementary File S1: All prediction accuracies (determination coefficient (R^2) and residual predictive deviations (RPD)) after four regression model analysis (random forest regression (RFR), partial least square regression (PLSR), least absolute shrinkage and selection operator regression (LassoR), and ridge regression (RidgeR)) based on two types of fatty acid measurements (g/100 g of milk and g/100 g fat), two ranges of wavebands (4000~400 cm^{-1} and 3017~2823 cm^{-1}/1805~1734 cm^{-1}), and five spectral pre-processing algorithms (first-order derivative (DER1), second-order derivative (DER2), multiple scattering correction (MSC), standard normal transform (SNV), and Savitzky–Golsy convolution smoothing (SG).

Author Contributions: Conceptualization, J.L., X.W. and D.S.; methodology, X.Z. and Y.S.; formal analysis, X.Z. and Y.S.; resources, Y.Z., G.C., G.X., Y.L. and Y.G.; data curation, K.C., F.Z., K.W. and M.Z.; writing—original draft preparation, X.Z., X.W. and Y.S.; writing—review and editing, X.W.; supervision, J.L. and X.W.; funding acquisition, J.L. and D.S. All authors have read and agreed to the published version of the manuscript.

Funding: This work was supported by Shandong Provincial Natural Science Foundation (ZR2021MC070), the National Key R&D Program of China (2021YFF1000701-06), Shandong Provincial Natural Science Foundation (ZR2020MC165), and the Earmarked Fund for CARS-36.

Institutional Review Board Statement: Not applicable.

Informed Consent Statement: Not applicable.

Data Availability Statement: Not applicable.

Acknowledgments: We thank the farm in Shandong Province, China, for providing the milk samples.

Conflicts of Interest: The authors declare no conflict of interest.

Sample Availability: Samples of the fatty acid quantification and the mid-infrared spectrum (MIRS) value are available from the authors.

References

1. Christie, W. The Composition and Structure of Milk Lipids. In *Developments in Dairy Chemistry—2*; Springer: Berlin/Heidelberg, Germany, 1983. [CrossRef]
2. Balaji, B.; Dehghan, M.; Mente, A.; Rangarajan, S.; Yusuf, S. Association of Dairy Consumption with Metabolic Syndrome, Hypertension and Diabetes in 147,812 Individuals from 21 Countries. *SSRN Electron. J.* **2019**. [CrossRef]
3. Haug, A.; Stmark, A.T.; Harstad, O.M. Bovine milk in human nutrition—A review. *Lipids Health Dis.* **2007**, *6*, 25. [CrossRef] [PubMed]
4. Soyeurt, H.; Dehareng, F.; Gengler, N.; McParland, S.; Wall, E.; Berry, D.P.; Coffey, M.; Dardenne, P. Mid-infrared prediction of bovine milk fatty acids across multiple breeds, production systems, and countries. *J. Dairy Sci.* **2011**, *94*, 1657–1667. [CrossRef]
5. Fleming, A.; Schenkel, F.S.; Chen, J.; Malchiodi, F.; Bonfatti, V.; Ali, R.A.; Mallard, B.; Corredig, M.; Miglior, F. Prediction of milk fatty acid content with mid-infrared spectroscopy in Canadian dairy cattle using differently distributed model development sets. *J. Dairy Sci.* **2017**, *100*, 5073–5081. [CrossRef] [PubMed]
6. Jensen, R.G. The composition of bovine milk lipids: January 1995 to December 2000. *J. Dairy Sci.* **2002**, *85*, 295–350. [CrossRef] [PubMed]
7. Stoop, W.M.; Bovenhuis, H.; Heck, J.M.L.; van Arendonk, J.A.M. Effect of lactation stage and energy status on milk fat composition of Holstein-Friesian cows. *J. Dairy Sci.* **2009**, *92*, 1469–1478. [CrossRef] [PubMed]
8. Wang, F.; Chen, M.; Luo, R.; Huang, G.; Wu, X.; Zheng, N.; Zhang, Y.; Wang, J. Fatty acid profiles of milk from Holstein cows, Jersey cows, buffalos, yaks, humans, goats, camels, and donkeys based on gas chromatography–mass spectrometry. *J. Dairy Sci.* **2021**, *105*, 1687–1700. [CrossRef]
9. Grelet, C.; Dardenne, P.; Soyeurt, H.; Fernandez, J.A.; Vanlierde, A.; Stevens, F.; Gengler, N.; Dehareng, F. Large-scale phenotyping in dairy sector using milk MIR spectra: Key factors affecting the quality of predictions. *Methods* **2021**, *186*, 97–111. [CrossRef]
10. Rovere, G.; de Los Campos, G.; Lock, A.; Worden, L.; Vazquez, A.I.; Lee, K.; Tempelman, R.J. Prediction of fatty acid composition using milk spectral data and its associations with various mid-infrared spectral regions in Michigan Holsteins. *J. Dairy Sci.* **2021**, *104*, 11242–11258. [CrossRef]
11. Liu, X.M.; Zhang, Y.; Zhou, Y.; Li, G.H.; Feng, X.S. Progress in Pretreatment and Analysis of Fatty Acids in Foods: An Update since 2012. *Sep. Purif. Rev.* **2019**, *50*, 203–222. [CrossRef]
12. Waktola, H.D.; Zeng, A.X.; Chin, S.-T.; Marriott, P.J. Advanced gas chromatography and mass spectrometry technologies for fatty acids and triacylglycerols analysis. *Trends Anal. Chem.* **2020**, *129*, 115957. [CrossRef]
13. Samková, E.; Špička, J.; Hanuš, O.; Roubal, P.; Pecová, L.; Hasoňová, L.; Smetana, P.; Klimešová, M.; Čítek, J. Comparison of Fatty Acid Proportions Determined by Mid-Infrared Spectroscopy and Gas Chromatography in Bulk and Individual Milk Samples. *Animals* **2020**, *10*, 1095. [CrossRef]
14. Lopez-Villalobos, N.; Spelman, R.; Melis, J.; Davis, S.; Berry, S.; Lehnert, K.; Sneddon, N.; Holroyd, S.; MacGibbon, A.; Snell, R. Genetic correlations of milk fatty acid contents predicted from milk mid-infrared spectra in New Zealand dairy cattle. *J. Dairy Sci.* **2020**, *103*, 7238–7248. [CrossRef]
15. Du, C.; Nan, L.; Yan, L.; Bu, Q.; Ren, X.; Zhang, Z.; Sabek, A.; Zhang, S. Genetic Analysis of Milk Production Traits and Mid-Infrared Spectra in Chinese Holstein Population. *Animals* **2020**, *10*, 139. [CrossRef]
16. Ferrand-Calmelsm, M.; Palhière, I.; Brochard, M.; Leray, O.; Astruc, J.; Aurel, M.; Barbey, S.; Bouvier, F.; Brunschwig, P.; Caillat, H.; et al. Prediction of fatty acid profiles in cow, ewe, and goat milk by mid-infrared spectrometry. *J. Dairy Sci.* **2014**, *97*, 17–35. [CrossRef]
17. Caredda, M.; Addis, M.; Ibba, I.; Leardi, R.; Scintu, M.F.; Piredda, G.; Sanna, G. Prediction of fatty acid content in sheep milk by Mid-Infrared spectrometry with a selection of wavelengths by Genetic Algorithms. *LWT Food Sci. Technol.* **2016**, *65*, 503–510. [CrossRef]
18. Caredda, M.; Addis, M.; Ibba, I.; Leardi, R.; Scintu, M.F.; Piredda, G.; Sanna, G. Building of prediction models by using Mid-Infrared spectroscopy and fatty acid profile to discriminate the geographical origin of sheep milk. *LWT Food Sci. Technol.* **2017**, *75*, 131–136. [CrossRef]
19. Nakayama, J.Y.; Ho, J.; Cartwright, E.; Simpson, R.; Hertzberg, V.S. Predictors of progression through the cascade of care to a cure for hepatitis C patients using decision trees and random forests. *Comput. Biol. Med.* **2021**, *134*, 104461. [CrossRef]
20. Wang, F.; Wang, Y.; Zhang, K.; Hu, M.; Weng, Q.; Zhang, H. Spatial heterogeneity modeling of water quality based on random forest regression and model interpretation. *Environ. Res.* **2021**, *202*, 111660. [CrossRef]
21. Soyeurt, H.; Dardenne, P.; Dehareng, F.; Lognay, G.; Veselko, D.; Marlier, M.; Bertozzi, C.; Mayeres, P.; Gengler, N. Estimating Fatty Acid Content in Cow Milk Using Mid-Infrared Spectrometry. *J. Dairy Sci.* **2006**, *89*, 3690–3695. [CrossRef]
22. Soyeurt, H.; Bastin, C.; Colinet, F.G.; Arnould, V.M.R.; Berry, D.P.; Wall, E.; Dehareng, F.; Nguyen, H.N.; Dardenne, P.; Schefers, J.; et al. Mid-infrared prediction of lactoferrin content in bovine milk: Potential indicator of mastitis. *Animal* **2012**, *6*, 1830–1838. [CrossRef]

23. Ho, P.N.; Marett, L.C.; Wales, W.J.; Axford, M.; Oakes, E.M.; Pryce, J.E. Predicting milk fatty acids and energy balance of dairy cows in Australia using milk mid-infrared spectroscopy. *Anim. Prod. Sci.* **2019**, *60*, 164. [CrossRef]
24. Tiplady, K.M.; Lopdell, T.J.; Littlejohn, M.D.; Garrick, D.J. The evolving role of Fourier-transform mid-infrared spectroscopy in genetic improvement of dairy cattle. *J. Anim. Sci. Biotechnol.* **2020**, *11*, 39. [CrossRef]
25. Soyeurt, H.; Dehareng, F.; Mayeres, P.; Bertozzi, C.; Gengler, N. Variation of delta9-desaturase activity in dairy cattle. *J. Dairy Sci.* **2008**, *91*, 3211–3224. [CrossRef]
26. Christie, W. Gas chromatography mass spectrometry methods for structural analysis of fatty acids. *Lipids* **1998**, *33*, 343–353. [CrossRef]
27. Maurice-van Eijndhoven, M.H.T.; Soyeurt, H.; Dehareng, F.; Calus, M. Validation of fatty acid predictions in milk using mid-infrared spectrometry across cattle breeds. *Animal* **2013**, *7*, 348–354. [CrossRef]
28. Coppa, M.; Ferlay, A.; Leroux, C.; Jestin, M.; Chilliard, Y.; Martin, B.; Andueza, D. Prediction of milk fatty acid composition by near infrared reflectance spectroscopy. *Int. Dairy J.* **2010**, *20*, 182–189. [CrossRef]
29. De Marchi, M.; Penasa, M.; Cecchinato, A.; Mele, M.; Secchiari, P.; Bittante, G. Effectiveness of mid-infrared spectroscopy to predict fatty acid composition of Brown Swiss bovine milk. *Animal* **2011**, *5*, 1653–1658. [CrossRef]
30. De Marchi, M.; Toffanin, V.; Cassandro, M.; Penasa, M. Invited review: Mid-infrared spectroscopy as phenotyping tool for milk traits. *J. Dairy Sci.* **2014**, *97*, 1171–1186. [CrossRef]
31. Zaalberg, R.M.; Poulsen, N.A.; Bovenhuis, H.; Sehested, J.; Larsen, L.B.; Buitenhuis, A.J. Genetic analysis on infrared-predicted milk minerals for Danish dairy cattle. *J. Dairy Sci.* **2021**, *104*, 8947–8958. [CrossRef]
32. Cecchinato, A.; Marchi, M.D.; Gallo, L.; Bittante, G.; Carnier, P. Mid-infrared spectroscopy predictions as indicator traits in breeding programs for enhanced coagulation properties of milk. *J. Dairy Sci.* **2009**, *92*, 5304–5313. [CrossRef] [PubMed]
33. Canaza-Cayo, A.W.; Alomar, D.; Quispe, E. Prediction of alpaca fibre quality by near-infrared reflectance spectroscopy. *Anim. Int. J. Anim. Biosci.* **2013**, *7*, 1219–1225. [CrossRef] [PubMed]
34. Pralle, R.S.; Weigel, K.W.; White, H.M. Predicting blood β-hydroxybutyrate using milk Fourier transform infrared spectrum, milk composition, and producer-reported variables with multiple linear regression, partial least squares regression, and artificial neural network. *J. Dairy Sci.* **2018**, *101*, 4378–4387. [CrossRef] [PubMed]
35. Rutten, M.J.M.; Bovenhuis, H.; Hettinga, K.A.; van Valenberg, H.J.F.; van Arendonk, J.A.M. Predicting bovine milk fat composition using infrared spectroscopy based on milk samples collected in winter and summer. *J. Dairy Sci.* **2009**, *92*, 6202–6209. [CrossRef]

molecules

MDPI

Article

Development of a Microsphere-Based Immunoassay Authenticating A2 Milk and Species Purity in the Milk Production Chain

Alexander J. W. Elferink [1], Deborah Entiriwaa [1,2], Paolo Bulgarelli [3], Nathalie G. E. Smits [1] and Jeroen Peters [1,*]

[1] Wageningen Food Safety Research, Wageningen University and Research, Akkermaalsbos 2, 6708 WB Wageningen, The Netherlands; alexander.elferink@wur.nl (A.J.W.E.); deborah.entiriwaa2017@my.ntu.ac.uk (D.E.); nathalie.smits@wur.nl (N.G.E.S.)

[2] School of Science & Technology, Nottingham Trent University, Clifton Lane, Nottingham NG11 8NS, UK

[3] Parmalat, Via delle Nazioni Unite 4, 43044 Collecchio, PR, Italy; p.bulgarelli@parmalat.net

* Correspondence: jeroen.peters@wur.nl; Tel.: +31-317-480579

Abstract: Processed milk and milk products produced from bovine milk, commonly contain β-casein A1 (βCA1) and β-casein A2 (βCA2). Since the presence of βCA1 is linked to milk intolerance and digestion problems, A2A2 milk, which only contains βCA2, is proposed as a healthier alternative. To support this health claim, the purity of A2A2-milk has to be guaranteed. In the presented study, a multiplex immunoassay, able to distinguish between βCA2 and βCA1, was developed and real-life applicability was shown on raw milk samples from genotyped A1A1, A1A2 and A2A2 cows. Because of its ability to discriminate between βCA2 and βCA1, this newly developed method was able to detect the addition of common bovine A1A2 milk to A2A2 milk, as low as 1%. Besides the detection of A2A2 milk purity, the developed assay can also be implemented as a rapid phenotyping method at dairy farms to replace the more invasive DNA-based screening. Additionally, the developed method was capable of detecting the addition of common bovine milk up to 1% in sheep, goat, buffalo, horse and donkey milk, which conforms to EU recommendations. In conclusion, a newly developed multiplex method capable of reliably detecting the dilution of A2A2 milk of multiple species, with common bovine milk up to 1%, is presented.

Keywords: milk; authenticity; purity; β-casein; A2; A1; species; microsphere immunoassay; multiplex; fraud

1. Introduction

Bovine milk is often consumed from infancy on, and throughout life, as it is a rich source of energy that contains many essential nutrients [1,2]. Bovine milk is rich in proteins, of which caseins make up approximately 80% of the total milk protein content. This heterogeneous group of caseins consists of four polymorphisms: αS1-, αS2-, β-, and κ-casein [3,4]. The most abundant group is αS1-casein at roughly 38%, followed by β-casein at 36%, κ-casein at 13% and αS2-casein at 10% [5]. The focus of this study is on the second most abundant casein, β-casein. The β-casein proteins consist of 209 amino acid (AA) residues and can be divided into twelve variants by their genetic polymorphisms; A1, A2, A3, B, C, D, E, F, G, H1, H2 and I [6–8]. However, only seven of these (A1, A2, A3, B, C, I, and E) have been detected in European cattle breeds, among which A1- and A2 β-caseins are the most prevalent variants [9,10]. The β-casein A2 (βCA2) genotype is considered the oldest genotype. Somewhere in time, a point mutation occurred in the βCA2 gene that caused the switch from proline to histidine at AA position 67, which led to the formation of the β-casein A1 (βCA1) variant (Figure 1). This single amino acid difference is the reason that βCA2 and βCA1 are much debated in current health claims [11].

During gastrointestinal digestion, proteases cleave βCA1 at positions 67 and 74 of the AA sequence, which releases a peptide of seven AAs called β-casomorphin 7 (BCM-7) [3]. BCM-7 is a significant bioactive peptide with strong morphine-like activity, which has been linked to adverse gastrointestinal effects, and diseases such as type 1 diabetes and coronary heart disease [12,13]. Furthermore, health issues due to milk consumption, such as irritable bowel syndrome, gastrointestinal dysfunction, and immune inflammation-related disorders were previously mainly allocated to lactose, but new insights provide evidence that casein proteins play an important role as well [14–16]. However, there are also doubts about these health effects, especially since substantial scientific research is lacking to fully support these health claims [14,17].

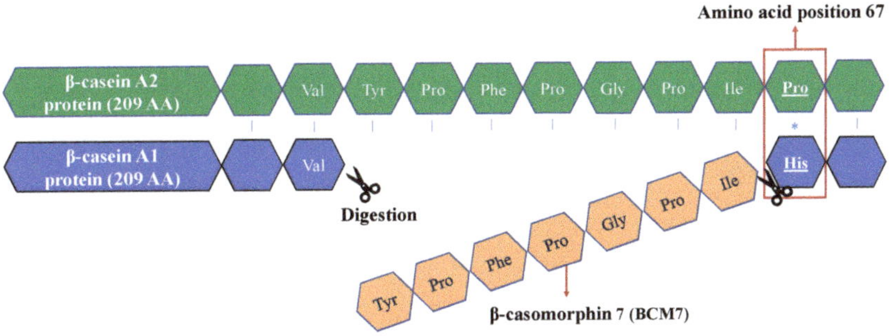

Figure 1. Schematic view of the partial amino acid sequences of the β-casein A2- and A1 protein. The red box and asterisk indicate the point mutation at position 67 (His in β-casein A1 instead of Pro in β-casein A2). The product of the gastrointestinal digestion of β-casein A1 is called β-casomorphin 7 (BCM-7) depicted in orange.

Upon gastrointestinal digestion of βCA2, the presence of proline at AA position 67 prevents the hydrolysis of the peptide bond, preventing the formation of BCM-7 [8]. The health risks assigned to the BCM-7 peptide, released upon digestion of βCA1, increased the demand for pure βCA2 milk products (e.g., bovine A2A2, goat, buffalo, or sheep milk) [18].

The Holstein breed (the most common dairy cow breed in Australia, Northern Europe, and the United States) carries both βCA2 and βCA1 alleles in approximately equal amounts, while Asian and African breeds are predominantly pure βCA2 [19]. In general, milk from common bovine dairy farms contains both βCA2 and βCA1 in various ratios and is referred to as common or A1A2 milk [20]. In recent years, the demand for pure βCA2 milk (A2A2 milk) has been growing as in many countries it is marketed as a functional food [21]. The difference in price and lack of suitable detection methods might make it attractive to adulterate expensive A2A2 milk types with cheaper, common A1A2 bovine milk [2,22]. Tampering with food has been conducted since ancient times; laws regarding adulteration even date back to Roman times [23,24]. More specifically, milk adulteration is a serious issue the dairy industry faces. Milk fraud causes major financial losses for the dairy industry; therefore, regulators, food producers, retailers, and consumers all have an interest in safeguarding milk products, ensuring they are safe, authentic, and of the highest quality [2]. Council Regulation No. 2018/150 of the commission of the European Communities requires producers to state the type of milk used for manufacturing cheese products [25]. Detecting adulteration prior to the fabrication of dairy products, requires a fast method [26]. Although no specific regulations are set for diluting pure A2A2 milk with common A1A2 milk, the detection limit (1%) of the test developed here seems applicable in regard to Council Regulation No. 2018/150, 'analysis and quality evaluation of milk and milk products eligible for public intervention and aid for private storage' that describes detection limits of 1 and 0.5% [25]. Logically, the intentional dilution of pure A2A2 milk with common A1A2 milk is only economically beneficial when more than a few percent is added. Fraudulent dilutions of pure A2A2 milk, consumed by people with an intolerance to milk products, can lead to the adverse effects assigned to the presence of βCA1 [27]. To track down fraudulent A2A2 milk dilutions, screening and analytical methods able to determine and distinguish between βCA2 and βCA1 need to be implemented throughout the milk supply chain. Several methods, to detect bovine milk as adulterant in milk from other species built on LC (Liquid Chromatography), ELISA (Enzyme Linked Immunosorbent Assay), PCR (Polymerase Chain Reaction), and PAGE (Polyacrylamide Gel Electrophoresis) were developed before [27–30]. However, none of the aforementioned studies were able to distinguish between bovine βCA2 and βCA1. Recently, Mayer et al. [31] described a method able to distinguish between βCA2 and βCA1 by using a restriction fragment length polymorphism (RFLP)-PCR. However, this is a time-consuming laborious method, not fit for high throughput sample analysis. Another method describes high-resolution melting (HRM) and rhAmp single nucleotide polymorphism (SNP) genotyping. These methods were able to identify the presence of the A1 allele in A2A2 milk samples; the HRM analysis showed a limit of detection of 10% (100 copies), whereas for rhAmp the limit was 2% (10 copies) [32]. The extensive extraction procedures, sample pretreatment and data processing, however, make these methods laborious and costly. Recently, Fourier transform infrared (FTIR) spectroscopy, coupled with multivariate analysis, was used to discriminate milk samples from Holstein-Friesian cows possessing different β-casein genetic variants by structural differences [33]. A commercial ELISA method from Biosensis is available, with a readout of results within 5 h. The test will only detect the presence of βCA1 in bovine milk samples and is not fit for testing milk from other species or phenotyping purposes. Moreover, if A1 levels are lower than 50%, additional pilot experiments will need to be conducted to determine optimal dilution levels. In the presented study, we developed a competitive microsphere-based multiplex immunoassay (cMIA) which is able to reliably distinguish between both the βCA2 and βCA1 proteins and was applied to detect the mixing of common A1A2 milk in pure A2A2 milk. Additionally, the cMIA was successfully applied to distinguish milk from pure A1A1, A2A2 and A1A2 bred cows. Next to indicating the βCA2/βCA1 ratio in milk, the developed cMIA is also capable of detecting adulteration of more expensive A2A2 milk types (e.g., goat, sheep, horse, donkey, and buffalo) with common bovine milk.

2. Results and Discussion

2.1. General

In the developed cMIA, the βCA1 and βCA2 proteins are coupled to the unique color-coded carboxylated paramagnetic microspheres. These coupled microspheres and two specific monoclonal antibodies (mAbs) against the βCA1 and βCA2 proteins are added to the diluted milk sample, all in one well. The free βCA1 and βCA2 proteins present in the milk samples will inhibit the binding of the mAbs to the βCA1 and βCA2 proteins on the microspheres (competitive assay). After a washing step, a secondary fluorescent reporter antibody is added for signal generation. This mixture containing two different microspheres will be captured on a planar by a magnet where the microspheres will be classified and their reporter signals (mean fluorescence intensities (MFIs)) quantified (Figure 2). Since it is a competitive assay format, assay responses will decrease as the β-casein concentrations increase. In other words, for the A1 assay the highest MFI values are obtained with a pure A2A2 milk sample, as it does not contain A1 β-caseins. This is inversed for the A2 assay. Here the highest MFI is obtained with a pure A1A1 milk sample as it contains no A2 β-caseins. To correct for MFI assay fluctuations caused by minor variations (e.g., temperature, new antibody reporter batch), the MFI of a specific fortified sample (B) will be divided by the MFI of a sample of the same type (or species) without fortification (B_0). This standardizes results obtained from different days.

2.2. Antibody Specificity

Both β-casein assays (βCA1- and βCA2) were initially developed as a single cMIA. Antibody dilutions for the single cMIAs were optimized to reach a signal of approximately

1000 median fluorescence intensity (MFI) for each assay. This resulted in antibody concentrations of 1.6 μg/mL (βCA1) and 0.47 μg/mL (βCA2), respectively. Next, the single cMIAs were combined to form a multiplex cMIA. Cross interactions were tested by incubating both β-casein-coupled microspheres with one single specific β-casein antibody. Initially, strong cross interactions were observed between the βCA1- and βCA2 assays. To overcome this problem, a range of buffers were tested (results not shown). The least cross interaction was observed when the standard PBS buffer (PBS with 0.02% Tween-20 and 0.1% BSA) was supplemented with 0.5 M NaCl and 0.25 M Na_2SO_4. Although the amino acid sequence of the βCA1- and βCA2 proteins only differ by one amino acid, both antibodies show high specificity for their corresponding βCA-protein coupled microspheres, as the level of cross interaction for both βCA1- and βCA2 assays is below 10% (Figure 3). For all consecutive multiplex measurements, both the β-casein-coupled microspheres and the βCA1- and βCA2 antibodies were combined in a single well. This means that for each analysis, two conjugated microspheres (βCA1- and βCA2) and two specific β-casein antibodies (βCA1 mAb- and βCA2 mAb) are present in one reaction well.

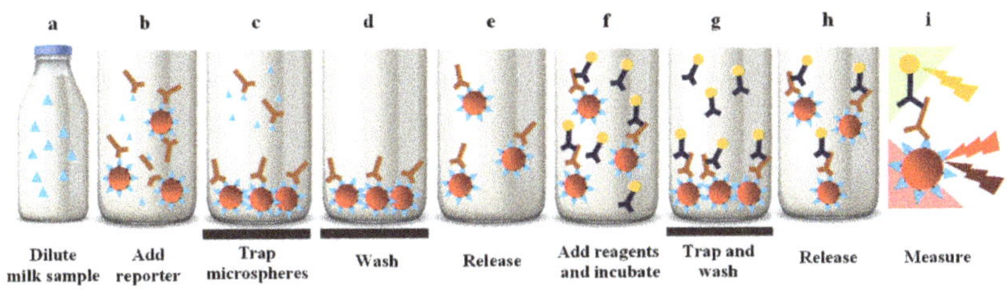

Figure 2. Sample analysis scheme for the competitive microsphere immunoassay. Milk samples are diluted (**a**), after which the β-casein-specific microspheres and β-casein-specific antibodies are added to the diluted sample (**b**); after incubation, the microspheres are trapped by a magnet (**c**) and the milk sample and unbound antibodies are removed (**d**). Next, the microspheres are released (**e**) and the GAC-RPE reporter antibody is added (**f**). After the second incubation step the microspheres are trapped and washed (**g**), released (**h**), and then measured in the planar array analyzer (**i**), where the microspheres are classified by a red LED illumination, after which the red and far-red fluorescence is recorded and the corresponding reporter signal is recorded after green LED illumination.

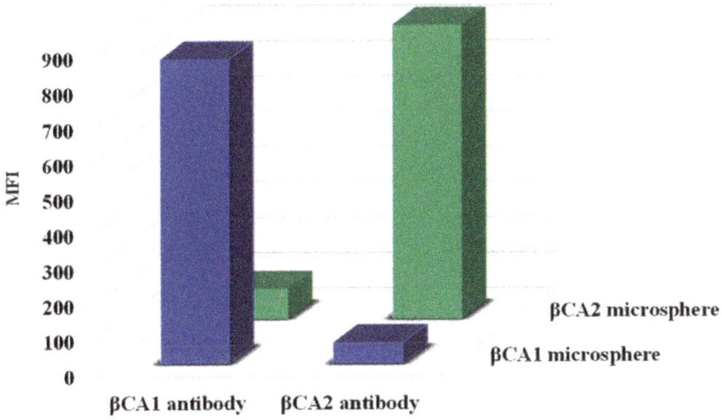

Figure 3. Specific and cross-interactive responses (MFI) of the β-casein assays in a competitive multiplex assay format where the multiplex microspheres are incubated with only a single antibody in the absence of milk samples. Blue bars are MFI values for βCA1 microspheres with βCA1- and βCA2 specific antibodies in buffered solution. Green bars are MFI values for βCA2 microspheres with βCA1- and βCA2 specific antibodies in buffered solution.

2.3. Determination of Optimal Milk Sample Dilution

β-caseins are present at high concentrations in milk (8.4 mg/mL) [34]. Because of these high concentrations, 10-fold milk sample dilutions were prepared, aiming for the optimal sample dilution that establishes the highest discrimination (i.e., the lowest cross-reactivity) in the βCA1- and βCA2 assays. Based on these results, as shown in Figure 4a,b, the highest discriminative power (lowest cross-reactivity) was obtained with a sample dilution of 10.000x. Therefore, this dilution was used in all subsequent experiments.

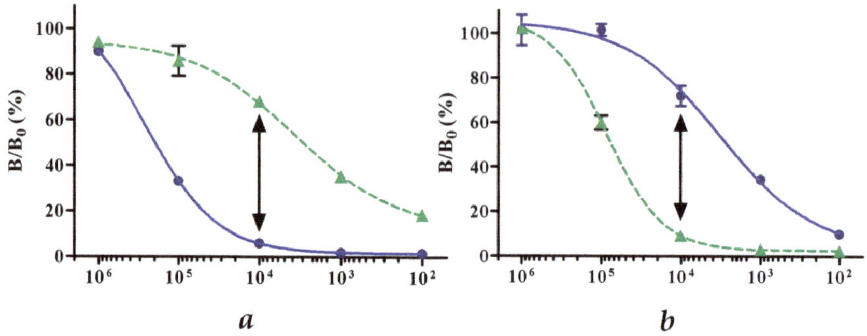

Figure 4. Absolute responses (B/B_0 (%)) for A1A1 milk (•) and A2A2 milk (▲) in 10-fold dilution series, in the βCA1 assay (**a**) and the βCA2 assay (**b**). The dilution generating the largest B/B_0 (%) difference between the A1A1 and A2A2 milk in both assays is indicated by the double-headed arrow.

2.4. Application of the cMIA to A1A1/A2A2 Mixed Raw Milk Samples

The pure A1A1, pure A2A2 and the mixed A1A2 samples were diluted 10,000 times and required no further sample treatment prior to analysis in the developed β-casein multiplex. The competitive format of the cMIA indicates a higher amount of a specific β-casein present in the sample by lower absolute signals. Figure 5 shows the dose-response results for various ratios of A1A1/A2A2 mixed milk samples. These results show that the signal in the βCA2 assay of the multiplex decreases when pure A1A1 milk is added. The decrease is significant up to 50%. After this the decrease flattens and becomes less prominent. However, from 50% onward the discriminative power comes from the βCA1 assay in the multiplex, as this assay shows a significant increase from this point onward. These results prove that the developed βCA2/βCA1 multiplex is a powerful tool for authenticating the purity of A2 milk. Based on the dose-response results, unknown samples analyzed by the cMIA can be semi-quantified for their A1A1 and A2A2 content. For the milk processing industry, this is an important tool to easily identify milk phenotype purity and track down milk adulteration. For the preparation of the dose-response ratio calibration mixtures (Figure 5), pure A1A1 milk was obtained from genetically selected cows that are kept for research purposes. In general, the natural occurrence of the A1A1 phenotype in offspring is low and therefore, pure A1A1 milk, in practice, will normally not be available for diluting A2A2 milk when considering additions at bulk tank level. Therefore, we decided to use common raw or pasteurized A1A2 milk for milk mixing, in all consecutive experiments.

2.5. Application to Supermarket- and β-Casein Phenotyped Milk Samples

A total of ten milk samples, consisting of seven raw bovine milk samples from cows, genotyped as A1A1, A1A2 and A2A2, and three pasteurized milk samples from the supermarket were analyzed by the cMIA (Figure 6).

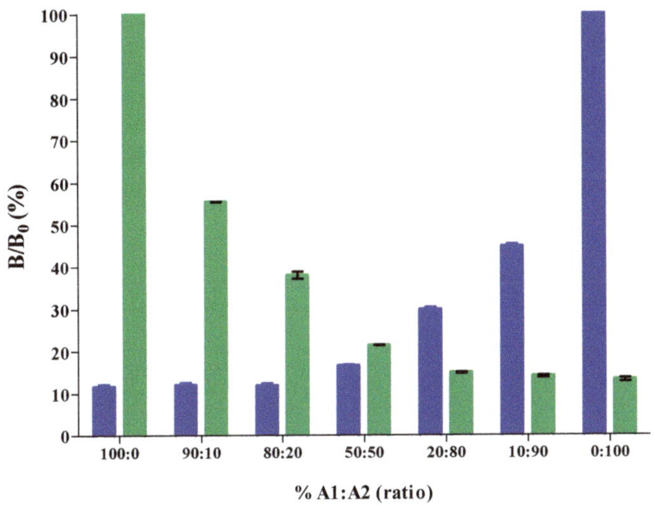

Figure 5. Dose-response results (B/B_0 (%)) for different ratios of A1A1/A2A2 milk mixtures in the competitive microsphere multiplex immunoassays detecting both βCA1 (■) and βCA2 (■).

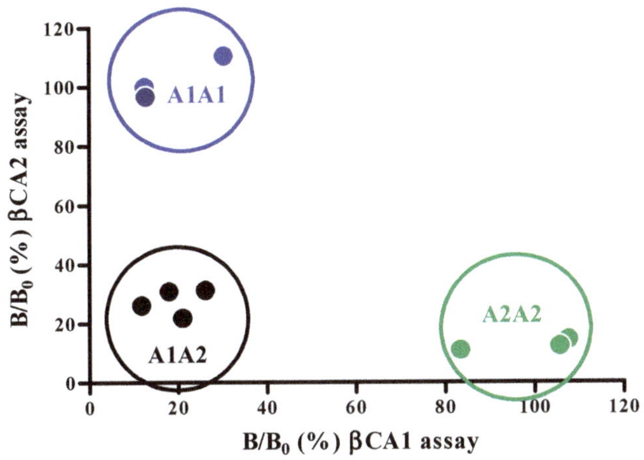

Figure 6. Phenotyping of raw milk samples collected from genotyped A1A1, A1A2 and A2A2 cows and pasteurized milk from supermarkets, by plotting the βCA1 and βCA2 responses against each other. Grouped in circles are the 3 phenotypes of milk tested, e.g., A1A1 milk (•), common milk A1A2 (•), and A2A2 milk (•).

Clear differentiation could be made between the three β-casein genotype groups (Figure 6). A1A1 milk samples only show inhibition in the βCA1 assay (*x*-axis), whereas the A2A2 milk only shows inhibition in the βCA2 assay (*y*-axis). Milk containing both β-caseins (A1A2) shows full inhibition for both assays. These results support the previous findings in the milk mixing experiments (Figure 5). Next to fraud detection in bulk milk, the cMIA can be used as a cow phenotyping tool, by simply analyzing the individual milk samples harvested from individual cows. This directly relates to the genotypes and therefore is a faster, less invasive, and less time-consuming method compared with most current DNA-based assays.

2.6. cMIA Applicability for Species' Milk Authenticity

To determine the applicability of the cMIA for determining species' milk authenticity, it was applied to eight A2A2 milk samples obtained from six different animal species (including cow) that were individually fortified with 1, 5, 10, 20 and 40% pasteurized common A1A2 bovine milk. For an enhanced interpretation of Figure 7, a break-down of the graph into individual samples is presented in the Supplementary Material (Figure S1).

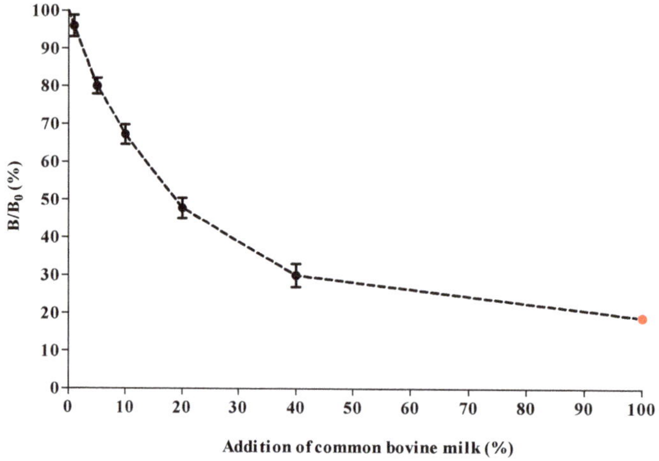

Figure 7. Combined average inhibition responses for the addition of pasteurized A1A2 bovine milk to pure A2A2 milk from sheep, horse, buffalo, goat, donkey, and cow. Each data point, including variation, is the mean value of the combined A2A2 milk samples originating from 6 different species. The data point for 100% addition of pasteurized A1A2 bovine milk (•) is a single measurement.

All samples were 10,000× diluted before measurement. The diluted A2A2 milk samples from different species showed a decrease in signal response in the βCA1 assay from 1% addition of A1A2 bovine milk onwards (Figure 7). This means that the βCA1 assay of the cMIA is fit for the purpose of detecting foreign milk at a relevant level. Moreover, measurement variations, shown in Figure 7, indicate that the cMIA is consistent in species' fraud detection, regardless of the species diluted. Upon addition of 20% common bovine milk, all species' samples already show a 50% inhibition of the absolute response. Matrix effects due to high fat and protein content can negatively influence milk testing [35,36]. For instance, the tested buffalo and sheep milk had a much higher fat content compared with milk from other species [37]. Nevertheless, these compositional variations did not affect the performance of the βCA1 assay, making it a robust method.

2.7. Initial In-house Interassay Repeatability Testing of Species' Milk Authentication

The total analysis time for 96 samples using the cMIA is about 2 h; this includes incubation, washing, and measurement. To test the inter-assay repeatability, all species' milk samples were measured in duplicate over three different days (n = 6). All the species' milk samples were measured without the addition of bovine milk (negative control, 0%) and with additions of 5 and 10% of raw bovine (A1A2) milk. The inter-assay results are presented in Table 1. The mean response values at 5% A1A2 milk addition range from 75 to 80, while at 10% addition of A1A2 milk, they range from 61 to 65. Besides the substantial inhibition of the absolute responses, the inter-assay variations are all well below 10% for the diluted samples, showing the high precision of the method for authenticating species' milk. These results correlate with those presented in Figure 7, wherein pasteurized A1A2 milk was used for dilution.

Table 1. Inter-assay repeatability for the βCA1 assay in the cMIA. Samples were Measured in Duplicate over 3 different days (n = 6) with dditions of 5- and 10% common raw bovine milk.

Species	5% Addition A1A2 Milk		10% Addition A1A2 Milk	
	Mean B/B_0 (%)	CV (%)	Mean B/B_0 (%)	CV (%)
Bovine (A2)	75	3	62	6
Sheep	81	8	65	5
Horse	77	7	62	5
Buffalo	77	5	61	5
Donkey	79	6	65	6
Goat	80	8	65	7

Mean B/B_0 (%): Average value of duplicate measurements performed in triplicate. CV (%): Coefficient of variation (n = 6).

3. Materials and Methods

3.1. Instrumentation

For readout of the results, a microsphere-dedicated planar array analyzer (MAGPIX) from Luminex (Austin, Texas, TX, USA) was used, operated with XPONENT software version 4.2. For the retention of the MagPlex microspheres during the antibody–microsphere coupling procedure, a DynaMag-2 (Invitrogen Dynal, Oslo, Norway) magnetic separator stand was applied. A Bühler TiMix 2 (Salm en Kipp, Breukelen, The Netherlands) was utilized for all microtiter plate incubation steps (at 450 rpm) and a REAX 2 overhead shaker (Heidolph, Schwabach, Germany) for microsphere coupling. A Vortex Genie 2 (Scientific Industries, New York, NY, USA) was used to mix samples.

3.2. Chemicals and Reagents

Drive fluid and the paramagnetic color-encoded microsphere sets 043 and 038 were obtained from Luminex. Cellstar 96-well culture microtiter plates were purchased from Greiner (Alphen a/d Rijn, the Netherlands). The βCA2 and βCA1 proteins and respective IgY antibodies were purchased from GeneTel Laboratories LLC (Madison, WI, USA). The secondary reporter antibody, Goat anti-chicken coupled to R-Phycoerythrin (GAC-RPE), was obtained from immuno Reagents Inc. (Raleigh, NC, USA). The 2-(N-morpholino)ethanesulfonicacid (MES), N-hydroxysulfosuccinimide (sulfo-NHS), 1-ethyl-3-(3-dimethylaminopropyl)carbodiimide (EDC) and bovine serum albumin (BSA) were purchased from Sigma-Aldrich (Zwijndrecht, the Netherlands). Phosphate buffered saline (PBS), and sodium phosphate monobasic ($NaH_2PO_4{\cdot}H_2O$) were obtained from Merck (Darmstadt, Germany).

3.3. Milk Samples

All the raw A1A1, A2A2, and A1A2 bovine milk samples were collected from genetically selected cows by Parmalat (Collecchio, Italy). Additionally, Parmalat supplied pasteurized bovine milk samples and raw milk samples from sheep and buffalo. Raw milk samples from a goat, horse, and donkey were obtained at specialized farms throughout the Netherlands. Additionally, pasteurized bovine and goat milk were purchased from local Dutch supermarkets. All milk samples were aliquoted and stored at −20 °C and defrosted before analysis.

3.4. Coupling of Microspheres

The paramagnetic microspheres utilized in this study have carboxylic groups on the surface, allowing the coupling of the βCA1 and βCA2 proteins (Figure 8). For this study, xMAP paramagnetic microspheres 043 and 038, respectively each having their own distinct color code, were implemented. The coupling procedure was carried out at room temperature using protein LoBind tubes. From each microsphere set, 200 µL was washed with 100 µL dH_2O. After washing, the microspheres were resuspended in 80 µL of $NaH_2PO_4{\cdot}H_2O$ pH 6.2 (NaPi) and activated by adding 10 µL 50 mM EDC and 10 µL of

50 mM sulfo-NHS. The microsphere suspensions were incubated for 20 min and gently mixed by pipetting, at 10 min intervals. After activation, each microsphere set was washed two times with 500 µL 50 mM MES pH 5 and subsequently resuspended in 500 µL MES buffer containing 0.1 mg/mL of the respective β-casein proteins. Covalent coupling of the β-caseins to the microspheres commenced for 2 h in the dark, keeping microspheres in suspension by rotation. After two washing steps, as described previously, the microspheres were blocked for 30 min by incubation with 500 µL PBS-TBN (PBS, 0.1% BSA, 0.05% NaN_3, 0.02% Tween-20) pH 7.4. After blocking, the microspheres were washed twice with 1 mL PBS-TBN. Finally, the coupled microspheres were resuspended in 200 µL PBS-TBN and stored in the dark at 4 °C until further use. To confirm successful microsphere coupling, the maximum Median Fluorescent Intensity (MFI) signal was determined by the addition of an excess specific antibody and secondary reporter antibody (GAC-RPE).

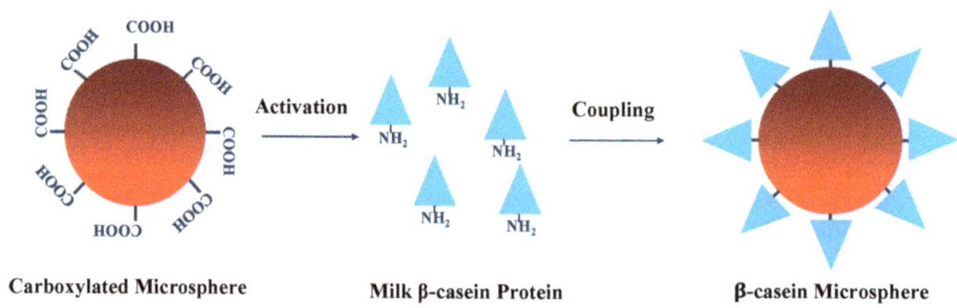

Figure 8. Graphical explanation of microsphere coupling procedure.

3.5. Multiplex Immunoassay

Before measurements, pure milk samples and mixed milk samples were 10,000× diluted. For the cMIA 100 µL of diluted milk sample (in 1× PBS, 0.1% BSA, 0.02% Tween, 0.25M Na_2SO_4, 0.5M NaCl), 10 µL microsphere suspension (1000 of each set), and 10 µL of prediluted βCA1 and βCA2 antibodies were added to a flat-bottom 96-well microplate (Figure 2). The mixture was incubated in the dark for 20 min on an orbital shaker. After incubation, the microspheres were trapped by a magnet, and the sample was washed twice with 100 µL of wash buffer (PBS, 0.02% Tween, 0.1% BSA). Next, the microspheres were released, and 100 µL of GAC-RPE reporter solution was added. After another 20 min incubation on the shaker, the microspheres were washed twice with 100 µL wash buffer and resuspended in PBST buffer to facilitate measurement in the planar array analyzer (Figure 2). MFI responses were normalized for daily variations, by dividing the responses of the fortified samples by the non-fortified pure milk samples (B/B_0).

4. Conclusions

The novel semi-high throughput multiplex method developed in this study is a useful tool for simple, robust, and reproducible authentication of raw and pasteurized A2A2 milk. The method is able to semi-quantitatively determine the ratio of βCA1 and βCA2 present in collected samples from milk tanks and end products and, additionally, enables the genotype prediction by analyzing milk samples collected from individual cows. With an analysis time of only 2 h for 96 samples, the presented multiplex method is 2.5 times faster than a commercially available ELISA test. Additionally, this ELISA needs several sample dilutions to fit in the dynamic range and requires strong alkaline solutions, all absent from our method. The current DNA-based methods need complicated DNA extractions before running the actual detection assay [31,32], whereas our method has an effortless, rapid, and simple sample preparation. The recently published FTIR spectroscopy method has the disadvantage of heavily relying on extensive centrifugation steps and a large database with specifications (e.g., environmental factors) to obtain reliable results [33]. Moreover, our method is, at the same time, a powerful tool to detect the unwanted presence of common bovine milk in milk from other species', from 1% onward. Therefore, it is relevant for real-life practice to detect carryover or mixing. In addition, the simple sample pretreatment and straightforward milk dilution step, shows good perspectives for future portable on-site detection. This assay format was already implemented for the on-site detection of antibiotics in chicken feathers, mycotoxins in beer samples and marine toxins in shellfish using standard pipettes, a small shaker, a handheld magnet for washing steps and electricity for operation. Future inter-laboratory and point-of-need validations based on end-user standards, should further prove the suitability for implementation of the developed method.

Supplementary Materials: The following supporting information can be downloaded at: https://www.mdpi.com/article/10.3390/molecules27103199/s1, Figure S1: Average inhibition responses for the addition of 0–40% pasteurized A1A2 bovine milk to pure A2A2 milk from sheep, horse, buffalo, goat, donkey and cow. The black line indicates the average response for the pure pasteurized A1A2 bovine milk, while the black dotted line indicates the average re-sponse of a pure A1A1 milk sample.

Author Contributions: A.J.W.E. and D.E. performed all experiments. P.B. studied and provided essential raw milk samples for this study. A.J.W.E. wrote the initial manuscript with substantial contributions from D.E. and P.B., while J.P. and A.J.W.E. were responsible for the methodology. J.P. and N.G.E.S. were responsible for the conceptualization of the research and writing and reviewing of the manuscript. All authors have read and agreed to the published version of the manuscript.

Funding: This work has received funding from the European Union's Horizon 2020 research and innovation program under grant agreement No 780839 (MOLOKO) and the Dutch Ministry of Agriculture, Nature and Food Quality (TKI-AF-EU-18027).

Institutional Review Board Statement: Not applicable.

Informed Consent Statement: Not applicable.

Data Availability Statement: Not applicable.

Acknowledgments: We are grateful to Willem Haasnoot for fruitful discussions, Rian Kraan for conducting initial experiments and Toine Bovee for critically reading the manuscript, all from Wageningen Food Safety Research. We also would like to thank Stefano Toffanin from the Institute for the Study of Nanostructured Materials—Italian National Research Council (CNR-ISMN), coordinator of the MOLOKO H2020 project where this research was conducted, for fruitful discussions.

Conflicts of Interest: The authors declare no conflict of interest.

Sample Availability: Milk samples used in this study are available from the authors upon request.

References

1. Clemens, R.A. Milk A1 and A2 peptides and diabetes. *Nestle Nutr. Workshop Ser. Pediatr. Program* **2011**, *67*, 187–195. [CrossRef] [PubMed]
2. Handford, C.E.; Campbell, K.; Elliott, C.T. Impacts of Milk Fraud on Food Safety and Nutrition with Special Emphasis on Developing Countries. *Compr. Rev. Food Sci. Food Saf.* **2016**, *15*, 130–142. [CrossRef] [PubMed]
3. Massella, E.; Piva, S.; Giacometti, F.; Liuzzo, G.; Zambrini, A.V.; Serraino, A. Evaluation of bovine beta casein polymorphism in two dairy farms located in northern Italy. *Ital. J. Food Saf.* **2017**, *6*, 6904. [CrossRef] [PubMed]
4. Fox, P.F.; Guinee, T.P.; Cogan, T.M.; McSweeney, P.L.H. *Fundamentals of Cheese Science*; 1 Online Resource (XV, 799 Pages 271 Illustrations, 778 Illustrations in Color; Springer: Boston, MA, USA, 2017. [CrossRef]
5. Kruif de, C.G.; Huppertz, T.; Urban, V.S.; Petukhov, A.V. Casein micelles and their internal structure. *Adv. Colloid Interface Sci.* **2012**, *171–172*, 36–52. [CrossRef] [PubMed]
6. Aschaffenburg, R. Inherited casein variants in cow's milk. *Nature* **1961**, *192*, 431–432. [CrossRef]
7. Peterson, R.F.; Kopfler, F.C. Detection of new types of β-casein by polyacrylamide gel electrophoresis at acid pH: A proposed nomenclature. *Biochem. Biophys. Res. Commun.* **1966**, *22*, 388–392. [CrossRef]
8. Firouzamandi, M.; Gholami, M.; Toloui, M.; Eshghi, D. Genetic Variation of β-Casein Gene Using AS-PCR and ARMS-PCR Techniques in Bovine Populations. *Russ. J. Genet.* **2018**, *54*, 1352–1357. [CrossRef]
9. Farrell, H.M., Jr.; Jimenez-Flores, R.; Bleck, G.T.; Brown, E.M.; Butler, J.E.; Creamer, L.K.; Hicks, C.L.; Hollar, C.M.; Ng-Kwai-Hang, K.F.; Swaisgood, H.E. Nomenclature of the proteins of cows' milk—Sixth revision. *J. Dairy Sci.* **2004**, *87*, 1641–1674. [CrossRef]
10. Sebastiani, C.; Arcangeli, C.; Ciullo, M.; Torricelli, M.; Cinti, G.; Fisichella, S.; Biagetti, M. Frequencies Evaluation of beta-Casein Gene Polymorphisms in Dairy Cows Reared in Central Italy. *Animals* **2020**, *10*, 252. [CrossRef]

11. Brooke-Taylor, S.; Dwyer, K.; Woodford, K.; Kost, N. Systematic Review of the Gastrointestinal Effects of A1 Compared with A2 beta-Casein. *Adv. Nutr.* **2017**, *8*, 739–748. [CrossRef]
12. Jianqin, S.; Leiming, X.; Lu, X.; Yelland, G.W.; Ni, J.; Clarke, A.J. Effects of milk containing only A2 beta casein versus milk containing both A1 and A2 beta casein proteins on gastrointestinal physiology, symptoms of discomfort, and cognitive behavior of people with self-reported intolerance to traditional cows' milk. *Nutr. J.* **2016**, *15*, 35. [CrossRef] [PubMed]
13. McLachlan, C.N.S. β-casein A1, ischaemic heart disease mortality, and other illnesses. *Med. Hypotheses* **2001**, *56*, 262–272. [CrossRef] [PubMed]
14. Daniloski, D.; Cunha, N.M.D.; McCarthy, N.A.; O'Callaghan, T.F.; McParland, S.; Vasiljevic, T. Health-related outcomes of genetic polymorphism of bovine β-casein variants: A systematic review of randomised controlled trials. *Trends Food Sci. Technol.* **2021**, *111*, 233–248. [CrossRef]
15. Ul Haq, M.R.; Kapila, R.; Sharma, R.; Saliganti, V.; Kapila, S. Comparative evaluation of cow beta-casein variants (A1/A2) consumption on Th-2-mediated inflammatory response in mouse gut. *Eur. J. Nutr.* **2014**, *53*, 1039–1049. [CrossRef]
16. Pal, S.; Woodford, K.; Kukuljan, S.; Ho, S. Milk Intolerance, Beta-Casein and Lactose. *Nutrients* **2015**, *7*, 7285–7297. [CrossRef]
17. Lambers, T.T.; Broeren, S.; Heck, J.; Bragt, M.; Huppertz, T. Processing affects beta-casomorphin peptide formation during simulated gastrointestinal digestion in both A1 and A2 milk. *Int. Dairy J.* **2021**, *121*, 105099. [CrossRef]
18. Truswell, A.S. The A2 milk case: A critical review. *Eur. J. Clin. Nutr.* **2005**, *59*, 623–631. [CrossRef]
19. Ng-Kwai-Hang, K.F.; Grosclaude, F. Genetic Polymorphism of Milk Proteins. In *Advanced Dairy Chemistry*; 1, Proteins; Fox, P.F., McSweeney, P.L.H., Eds.; Springer: Boston, MA, USA, 2003; pp. 739–816. [CrossRef]
20. Kaskous, S. A1- and A2-Milk and Their Effect on Human Health. *J. Food Eng. Technol.* **2020**, *9*, 15–21. [CrossRef]
21. Bentivoglio, D.; Finco, A.; Bucci, G.; Staffolani, G. Is There a Promising Market for the A2 Milk? Analysis of Italian Consumer Preferences. *Sustainability* **2020**, *12*, 6763. [CrossRef]
22. Kamthania, M. Milk Adultration: Methods of Detection & Remedial Measures. *Int. J. Eng. Tech. Res.* **2014**, *1*, 2321–2869.
23. Armstrong, D.J. Food chemistry and U.S. food regulations. *J. Agric. Food Chem.* **2009**, *57*, 8180–8186. [CrossRef] [PubMed]
24. Spink, J.; Moyer, D.C. Defining the public health threat of food fraud. *J. Food Sci.* **2011**, *76*, R157–R163. [CrossRef] [PubMed]
25. European Commission. Commission Implementing Regulation (EU) 2018/150 of 30 January 2018 Amending Implementing Regulation (EU) 2016/1240 as Regards Methods for the Analysis and Quality Evaluation of Milk and Milk Products Eligible for Public Intervention and Aid for Private Storage (*OJ* 2018, p. 26). Available online: https://eur-lex.europa.eu/legal-content/EN/TXT/?uri=CELEX%3A32018R0150 (accessed on 29 March 2022).
26. Haasnoot, W.; Smits, N.G.; Kemmers-Voncken, A.E.; Bremer, M.G. Fast biosensor immunoassays for the detection of cows' milk in the milk of ewes and goats. *J. Dairy Res.* **2004**, *71*, 322–329. [CrossRef] [PubMed]
27. Hurley, I.P.; Coleman, R.C.; Ireland, H.E.; Williams, J.H.H. Measurement of Bovine IgG by Indirect Competitive ELISA as a Means of Detecting Milk Adulteration. *J. Dairy Sci.* **2004**, *87*, 543–549. [CrossRef]
28. Chen, R.K.; Chang, L.W.; Chung, Y.Y.; Lee, M.H.; Ling, Y.C. Quantification of cow milk adulteration in goat milk using high-performance liquid chromatography with electrospray ionization mass spectrometry. *Rapid Commun. Mass Spectrom.* **2004**, *18*, 1167–1171. [CrossRef]
29. Bania, J.; Ugorski, M.; Polanowski, A.; Adamczyk, E. Application of polymerase chain reaction for detection of goats' milk adulteration by milk of cow. *J. Dairy Res.* **2001**, *68*, 333–336. [CrossRef]
30. Lopez-Calleja, I.; Gonzalez, I.; Fajardo, V.; Rodriguez, M.A.; Hernandez, P.E.; Garcia, T.; Martin, R. Rapid detection of cows' milk in sheeps' and goats' milk by a species-specific polymerase chain reaction technique. *J. Dairy Sci.* **2004**, *87*, 2839–2845. [CrossRef]
31. Mayer, H.K.; Lenz, K.; Halbauer, E.M. "A2 milk" authentication using isoelectric focusing and different PCR techniques. *Food Res. Int.* **2021**, *147*, 110523. [CrossRef]
32. Giglioti, R.; Gutmanis, G.; Katiki, L.M.; Okino, C.H.; de Sena Oliveira, M.C.; Vercesi Filho, A.E. New high-sensitive rhAmp method for A1 allele detection in A2 milk samples. *Food Chem.* **2020**, *313*, 126167. [CrossRef]
33. Daniloski, D.; McCarthy, N.A.; O'Callaghan, T.F.; Vasiljevic, T. Authentication of β-casein milk phenotypes using FTIR spectroscopy. *Int. Dairy J.* **2022**, *129*, 105350. [CrossRef]
34. Ng-Kwai-Hang, K.F.; Hayes, J.F.; Moxley, J.E.; Monardes, H.G. Variation in milk protein concentrations associated with genetic polymorphism and environmental factors. *J. Dairy Sci.* **1987**, *70*, 563–570. [CrossRef]
35. Selby, C. Interference in immunoassay. *Ann. Clin. Biochem.* **1999**, *36*, 704–721. [CrossRef] [PubMed]
36. Andrew, S.M. Effect of fat and protein content of milk from individual cows on the specificity rates of antibiotic residue screening tests. *J. Dairy Sci.* **2000**, *83*, 2992–2997. [CrossRef]
37. Gantner, V.; Mijić, P.; Baban, M.; Škrtić, Z.; Turalija, A. The overall and fat composition of milk of various species. *Mljekarstvo* **2015**, *65*, 223–231. [CrossRef]

molecules

MDPI

Article

Content and Nutritional Evaluation of Zinc in PDO and Traditional Italian Cheeses

Pamela Manzi *, Maria Gabriella Di Costanzo and Mena Ritota

Consiglio per la Ricerca in Agricoltura e L'analisi Dell'economia Agraria, Centro di Ricerca Alimenti e Nutrizione, Via Ardeatina 546, 00178 Rome, Italy; mariagabriella.dicostanzo@crea.gov.it (M.G.D.C.); mena.ritota@crea.gov.it (M.R.)
* Correspondence: pamela.manzi@crea.gov.it; Tel.: +39-065-149-4499

Abstract: Zinc is an essential mineral which plays a key role in several important biological processes in the human body. The determination of its level in food matrices can contribute to the food quality characterization and to the adequacy of the diet. Animal food products generally have a higher zinc content compared to vegetables. Among them, dairy products consumption can provide a great contribution to the zinc reference intakes. In this study, different Italian cheeses (38 Protected Denomination of Origin and 9 Traditional) were evaluated for their zinc content. Cow cheeses generally showed the highest zinc content (1.83–7.75 mg/100 g cheese), followed by sheep cheeses (1.34–3.69 mg/100 g), and cheeses from mixed milk (0.39–4.54 mg/100 g). The only cheese from buffalo milk (Mozzarella di Bufala Campana PDO) showed a zinc content of 2.14 mg/100 g. The great variability in the zinc content observed among the samples is the result of the influence of several factors, such as the feeding system, the species (cow, sheep, goat, and buffalo), and the cheese-making. Most of the samples resulted in a great contribution (>10%) to the zinc Daily Reference Intake set by EU (10 mg/day), with only two samples contributing to less than 4%.

Keywords: zinc; PDO and traditional cheeses; nutritional evaluation; dietary reference intake

1. Introduction

Zinc, a stable divalent cation, is one of the most abundant elements in the world. In the human body, zinc is the second micronutrient after iron, and it plays a key role in several important biological processes [1]. According to numerous authors [1–5], three different functions can be ascribed to zinc: catalytic, structural, and regulatory roles. Many enzymes require zinc as catalyst, such as oxidoreductases, transferases, hydrolases, lysases, isomerases, and ligases [6], where zinc may activate these enzymes by serving as an electron acceptor. The zinc structural role is related to its ability to promote the folding of proteins into three-dimensional configurations, involving above all cysteine and histidine [2], and because it is considered of critical importance in maintaining the structure of metalloproteins [4]. Finally, it has recently been shown that zinc plays an essential role in polynucleotide transcription and translation, and thus in the processes of genetic expression (such as expression of the metallothionein gene, apoptosis, and synaptic signaling) [5].

As zinc is involved in several functions, its content in the human body ranges from 1.5 to 2.6 g in adults, with bone and skeletal muscle accounting for the highest amount (about 86%), followed by skin (4.2%), and liver (3.4%) [3]. However, zinc is present in all organs, tissues, fluids, and secretions of the body. There is no storage compartment for zinc in the body, and the zinc losses (through feces, urine, sweat, menstrual flow, loss of hair, nails, and desquamated skin) must be replaced by dietary intake. The maintenance of zinc homeostasis is essential for survival, hence, regulations in zinc absorption and excretion are useful to maintain zinc homeostasis [7].

If the supply of dietary zinc is insufficient, biochemical abnormalities and clinical signs may develop, including pregnancy outcome, physical growth, susceptibility to infection, and neurobehavioral development [2,6,8]. This is why infants, children, pregnant, and lactating women are at greater risk of zinc depletion compared to adults.

For this reason, great efforts have been done to estimate the global prevalence of zinc deficiency using accurate and easily measurable indicators of zinc status. A working group made up by the World Health Organization (WHO), the United Nations International Children's Emergency Fund (UNICEF), the International Atomic Energy Agency (IAEA), and the International Zinc Nutrition Consultative Group (IZiNCG) recommend the blood plasma or the serum zinc concentration as biomarker of the zinc deficiency risk in populations [9].

Many food items are a valuable source of zinc [2]: animal foods, such as liver, meat, seafood, eggs, and dairy products, have higher zinc contents compared to vegetables (beans, cereal, tubers, vegetables, and fruits). However, meal studies based on the investigations of the real utilization of dietary zinc showed the presence of different factors affecting zinc absorption [10,11]. Among them, the ability of myo-inositol hexaphosphate (also known as phytic acid) to interfere with the zinc absorption is the most studied [4, 12–17]. Phytic acid chelates cations, forming insoluble complexes with minerals that, in this way, cannot be digested or absorbed by humans [18]. However, among the phosphorylated forms of inositol phosphate, the hexaphosphate and pentaphosphate forms inhibited zinc absorption, while the tetraphosphate and triphosphate forms had no significant effect [16]. It is worth highlighting that dietary proteins can modify the effects of phytate [14], since animal proteins have been shown to improve zinc absorption from a phytate-containing diet [19].

Many authors suggest [6,16] the phytates/zinc molar ratio in meals or diets as a useful indicator of the phytates degree in depressing zinc absorption. In a meal with a molar ratio above 15, for example, the zinc absorption is typically less than 15% [19].

Plant-based diets, containing high phytate amounts, provide less available zinc than meat-based diets [15], while animal source foods, such as dairy products, do not contain phytate and therefore have a phytate/zinc molar ratio equivalent to zero. In this regard, some authors [11] showed that zinc bioavailability in soy-based beverage was much lower (≈20%) compared to bovine milk, where the zinc bioaccessible fractions ranged from 59% to 83%, just because of the presence of phytic acid in the soy-based beverage samples. The authors [11] also observed no significant differences among the samples of whole, semi-skimmed and skimmed milk, showing that the lipid content does not seem to affect the zinc bioaccessibility in bovine milk.

The role of calcium in affecting zinc availability is controversial: it does not seem to have any direct effect on zinc absorption [20] but, in the presence of phytate, calcium may increase the phytate inhibitory capacity. Negative effects of some ions, such as iron, on zinc absorption have been reported, while soluble organic substances of low molecular mass, such as amino acids or citric acid, facilitate zinc absorption [3]. Additionally, caseinophosphopeptides (CPPs) seem to increase zinc absorption [21,22]. They can be released by enzymatic digestion of dairy products or during cheese-making [23]. CPPs contain three serine phosphate clusters followed by two glutamic acid residues, and they have a strong capacity to fix divalent cations, such as calcium, iron or zinc, by forming soluble organophosphate salts [21,22].

Hence, since zinc is an essential trace element with a great nutritional importance, the determination of its level in food matrices can contribute to the characterization of the food quality and to the adequacy of the diet. However, it can also result in being toxic for humans when taken in excess [24,25], so much so that the Provisional Maximum Tolerable Daily Intake (PMTDI) of zinc has been fixed to 1.00 mg/kg body weight [26]. Therefore, its determination in food products is of fundamental importance.

In this context, the aim of this work is to update food composition databases with reliable and representative data, with a focus on the zinc content in Protected Denomination of Origin (PDO) and Traditional Italian cheeses. PDO products have the strongest links to the place in which they are made, and every part of the production, processing and preparation process must take place in the specific region. Traditional cheeses are, instead, typical and local products whose processing, preservation, and seasoning methods have been consolidated over time almost since 25 years according to traditional Italian Regulation [27]. Since PDO and Traditional Italian cheeses are food items largely consumed in the European Union (EU), this study also aims to provide the nutritional adequacy of zinc in these cheeses, according to the UE regulation and European Food Safety Authority (EFSA) Population Reference Intakes for zinc [13,28].

2. Results and Discussion

2.1. Zinc Levels in PDO and Traditional Italian Cheeses

Table 1 shows the zinc content of PDO and Traditional cow cheese samples (expressed in wet weight): the mean data range from 2.28 mg/100 g (Taleggio) and 7.75 mg/100 g (Piave Vecchio) in PDO cheeses, while among Traditional cheeses Giuncata shows the lowest zinc amount (1.83 mg/100 g) and Provola delle Madonie (18 months ripening) has the highest zinc content (7.18 mg/100 g). The differences in these cheese samples partially reflect their water contents [29].

Table 1. Zinc content in PDO and Traditional Italian cow cheeses (mg/100 g).

Samples	Samples	* N	Min	Max	Mean	Std Dev
PDO	Asiago	5	4.22	5.25	4.61	0.44
PDO	Asiago Pressato	10	3.47	5.39	4.08	0.66
PDO	Bitto	2	4.06	4.90	4.48	0.59
PDO	Caciocavallo Silano	2	4.38	4.59	4.48	0.15
PDO	Casera	2	4.81	5.10	4.93	0.15
PDO	Fontina	6	3.49	4.69	3.95	0.50
PDO	Formai de Mut	2	4.59	4.84	4.72	0.18
PDO	Gorgonzola	8	2.43	3.35	2.94	0.29
PDO	Grana Padano (>16 months)	5	5.33	5.76	5.48	0.18
PDO	Montasio	4	4.16	5.05	4.63	0.39
PDO	Monteveronese	2	3.77	4.36	4.07	0.42
PDO	Monteveronese d'allevo	3	4.02	4.35	4.29	0.25
PDO	Nostrano Valtrompia					
	Alpine pasture	4	3.94	5.27	4.65	0.71
	indoor	4	4.95	5.06	5.01	0.05
PDO	Parmigiano Reggiano (>18 months)	5	5.01	5.54	5.29	0.19
PDO	Piave Vecchio	2	7.53	7.96	7.75	0.31
PDO	Provolone Valpadana	5	3.67	4.51	4.17	0.31
PDO	Puzzone di Moena	2	4.07	4.35	4.21	0.20
PDO	Quartirolo Lombardo	2	2.99	3.84	3.42	0.60
PDO	Ragusano	4	4.45	5.52	4.85	0.47
PDO	Spressa delle Giudicarie	2	5.73	5.94	5.83	0.15
PDO	Stelvio	2	3.92	4.19	4.05	0.19
PDO	Taleggio	5	2.04	2.42	2.28	0.15
PDO	Toma Piemontese	5	3.20	4.28	3.95	0.43
Traditional	Bagoss					
	Alpine pasture	6	5.25	7.98	6.23	1.36
	indoor	4	5.77	6.06	5.90	0.14
Traditional	Caciotta di Raiano	3	3.53	3.58	3.55	0.03
Traditional	Formaggella della Val Trompia					
	Alpine pasture	4	4.14	4.72	4.45	0.28
	indoor	4	3.18	3.58	3.39	0.17
Traditional	Giuncata abruzzese Vaccina	2	1.78	1.88	1.83	0.07
Traditional	Provola delle Madonie					
	Unripened	6	4.52	5.25	4.87	0.37
	4 months	6	5.70	6.86	6.26	0.60
	6 months	6	5.75	8.65	7.18	1.51
	Smoked	6	4.43	5.26	4.79	0.37

* N = number of cheeses obtained from different farmers.

These data are quite similar to those reported by Gambelli et al. [30], especially for the zinc content of Gorgonzola, Caciocavallo, Provolone, and Grana. Kedzierska-Matysek et al. [31] reported similar values for Grana Padano and Fontina zinc contents, 5.00 and 3.49 mg/100 g, respectively. Bontempo et al. [32], in the evaluation of the mineral contents of several Italian cheeses, reported comparable Asiago zinc value (6.50 mg/100 g on a dry weight), while the zinc levels of Fontina, Montasio, Puzzone, Spressa, and Toma were slightly lower than those obtained for the same cheeses in this study. Additionally, Manuelian et al. [33] reported lower zinc levels for Asiago, Grana Padano, Montasio, Parmigiano Reggiano, Piave, Fontina, Provolone, Gorgonzola, and Taleggio, compared to the results obtained in this work, and the data reported by some authors [34,35] for the zinc content of Quartirolo cheese were lower than that obtained in this study.

The zinc content of Parmigiano Reggiano (>18 months ripening) is in the same order of magnitude of the data reported by Kedzierska-Matysek et al. [31], where the zinc values varied between 4.25 and 4.53 mg/100 g. The zinc amount in Giuncata vaccina abruzzese, a soft dairy product for immediate consumption, is comparable with the results obtained by Ianni et al. [36], who reported a zinc value of 2.07 and 1.99 mg/100 g for the same cheese, while the zinc level reported by Barone et al. [37] was lower (0.40 mg/100 g). The zinc contents of Caciocavallo reported by Ianni et al. [38], ranging from 4.36 to 4.20 mg/100 g, are in the same order of magnitude of the data reported for the same cheese in this study and by Gambelli et al. [30].

Table 1 also shows the same type of cheeses produced with milk where cows were bred in indoor or grazed the alpine pasture: this is the case of Nostrano Valtrompia (PDO cheese), Bagoss and Formaggella della Val Trompia (Traditional cheeses). The zinc content decreases ($p < 0.05$) in the cheese samples made with milk from indoor feeding only in Formaggella della Val Trompia cheeses. These results are confirmed even when performing statistical analyses on a dry basis (data not shown). The influence of the feeding system, as well as the season, on milk composition is already known [39]. However, finding a univocal relation between the feeding system and the zinc content in cheese is very difficult due to the influence of several factors: (i) there is a great variability in the mineral contents of mountain pasture plants [40]; (ii) the intensity of grazing could affect mineral levels of the forage [41]; zinc values and its threshold values in soils are not easy to evaluate due to several variables involved [42]; and zinc content in milk could be affected by zinc supplementation in the feeding system [43,44], even if zinc levels in cheese depend on cheese-making processes [45–48], with curdling and salting being the two phases mainly influencing cheese mineral composition [34]. In confirmation of all these variables simultaneously affecting the zinc content of cheese, there are the studies of Coni et al. [34] and Lucas et al. [49], according to which no influence of the milk season and feeding practices, respectively, on the zinc content of cow cheeses was observed.

Among Traditional Italian cheeses made from cow milk, there seems to be a ripening influence on the zinc content: this is the case of Provola delle Madonie, a pasta filata cheese, in which an increasing trend of the zinc level from unripened to 16 months ripening is observed. Actually, since the same cheeses come from the same farm, the differences are only due to the decreasing water content with ripening, so much so that the significance differences disappear when performing statistical analysis on a dry basis (data not shown). Table 2 shows the zinc content in PDO and Traditional sheep cheeses (expressed in wet weight). According to some authors sheep cheeses are a good source of zinc [50]. In PDO sheep samples, the mean values of zinc range from 2.72 mg/100 g (Formaggio di Fossa di Sogliano) to 3.69 mg/100 g (Canestrato Pugliese), while in Traditional sheep cheeses the zinc values vary from 1.34 (Giuncatella abruzzese) to 3.56 mg/100 g (Pecorino d'Abruzzo, 6 months ripening). In PDO sheep samples, the mean values of zinc range from 2.72 mg/100 g (Formaggio di Fossa di Sogliano) to 3.69 mg/100 g (Canestrato Pugliese), while in Traditional sheep cheeses the zinc values vary from 1.34 (Giuncatella abruzzese) to 3.56 mg/100 g (Pecorino d'Abruzzo, 6 months ripening).

Table 2. Zinc content in PDO and Traditional Italian sheep cheeses (mg/100 g).

Samples		* N	Min	Max	Mean	Std Dev
PDO	Canestrato Pugliese	2	3.49	3.89	3.69	0.19
PDO	Fiore Sardo	5	2.20	3.74	3.25	0.61
PDO	Formaggio di Fossa di Sogliano	2	2.67	2.77	2.72	0.07
PDO	Pecorino Romano	6	3.15	4.18	3.66	0.38
PDO	Pecorino Sardo	4	2.62	4.11	3.60	0.67
PDO	Pecorino Toscano	6	2.31	3.54	2.81	0.50
Traditional	Giuncatella abruzzese	5	0.69	1.77	1.34	0.58
Traditional	Pecorino d'Abruzzo					
	Primo Sale	8	2.38	3.24	2.89	0.29
	2 months	3	3.06	3.14	3.10	0.04
	3 months	2	2.82	2.88	2.85	0.04
	4 months	10	2.34	3.43	2.79	0.39
	6 months	3	3.41	3.70	3.56	0.15
	7 months	2	2.08	2.27	2.18	0.14
	18 months	2	3.46	3.53	3.50	0.05

* N = number of cheeses obtained from different farmers.

Pecorino d'Abruzzo, a largely consumed cheese in Italy, shows on average 2.98 mg/100 g of zinc, and this value is comparable with the results obtained by Gambelli et al. [30] who reported a mean value of 3.37 mg/100 g for Pecorino-type cheese. Kedzierska-Matysek et al. [31], instead, reported a lower zinc content in Pecorino Toscano compared to the same cheese analysed in this study (1.77 vs. 2.81 mg/100 g).

Table 3 reports the zinc content in PDO cheeses from buffalo and mixed milk (expressed in wet weight). Mozzarella di Bufala Campana PDO shows 2.14 mg/100 g of zinc, with a similar value reported by Gambelli et al. [30] and Bontempo et al. [51] for the same cheese, while Manuelian et al. [33] observed a lower zinc content in Mozzarella di Bufala Campana PDO (1.34 mg/100 g).

In cheeses from mixed milk, instead, the zinc value ranges from 2.07 to 4.54 mg/100 g (Murazzano and Bra duro, hard type, respectively), excluding Robiola di Roccaverano and Ricotta Romana PDO, where the zinc values are respectively 0.39 and 0.36 mg/100 g.

Ricotta Romana PDO is not really a cheese, but a dairy product obtained from whey (sheep whey about 85%). As the soluble fraction of sheep milk the zinc amount is 8.4% [52], the level of this element observed in Ricotta Romana is quite low (0.36 mg/100 g). For the same reason, the data reported by Coni et al. [53] revealed that the zinc content in Ricotta-type cheese was very low (0.16 mg/100 g). The result obtained in this study is also comparable with the data of Gambelli et al. [30], where the zinc content in Ricotta-type cheese (from cow and sheep milk) is 0.35 mg/100 g.

In this study cow milk cheeses result in higher ($p < 0.05$) average zinc content (4.61 mg/100 g) compared to sheep, mixed, and buffalo cheeses (3.00, 2.60, and 2.14 mg/100 g, respectively) as reported in Figure 1. The high variability in mixed cheeses is due to the great heterogeneity of the samples, made with different technological processes and with different percentages of milk from various animal species.

Table 3. Zinc content in PDO Italian cheeses from buffalo milk, from mixed milk, and from sheep whey (mg/100 g).

Samples		* N	Min	Max	Mean	Std Dev	Milk Origin
PDO	Mozzarella di Bufala Campana	3	1.93	2.47	2.14	0.29	buffalo
PDO	Bra duro (*hard type*)	2	4.12	4.97	4.54	0.60	cow (small amounts goat or sheep milk)
PDO	Bra Tenero (*soft type*)	3	3.58	3.85	3.69	0.14	cow (small amounts goat or sheep milk)
PDO	Castelmagno	3	1.73	3.35	2.64	0.83	cow (small amounts goat or sheep milk)
PDO	Raschera	2	3.74	4.05	3.90	0.22	cow (small amounts goat or sheep milk)
PDO	Caciotta d'Urbino	2	3.19	3.19	3.19	0.09	sheep (70–80%) + cow milk (30–20%)
PDO	Murazzano	2	1.99	2.13	2.07	0.07	sheep or sheep (>60%) + cow (<40%) milk
PDO	Robiola di Roccaverano	2	0.37	0.41	0.39	0.03	goat or goat (50%) + cow/sheep (50%) milk
PDO	Ricotta Romana	2	0.27	0.45	0.36	0.13	sheep whey + sheep milk (max 15%)

* N = number of cheeses obtained from different farmers.

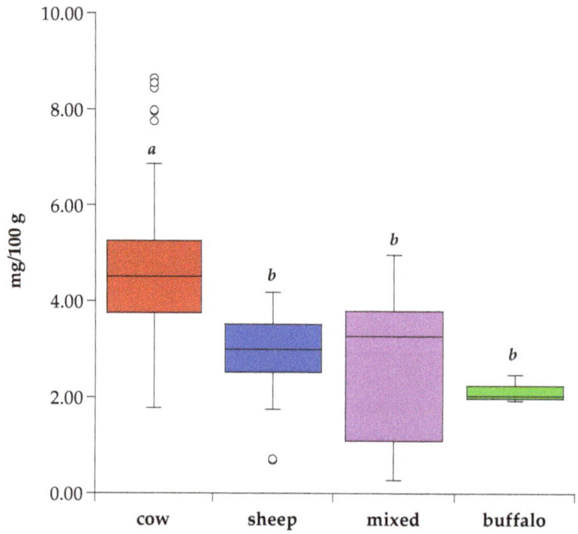

Figure 1. Box plots of the zinc level in cow, sheep, mixed, and buffalo milk cheeses. Significant differences ($p < 0.05$) are identified with different letters (*a*, *b*).

Additionally, Manuelian et al. [33] evaluated the zinc content in cheeses with milk from different species reporting the following trend: cow > sheep > buffalo >> goat. Similarly, Martin-Hernandez et al. [54] observed a lower zinc content in goat cheeses than cow and sheep cheeses. Since the differences in the mineral composition of milk from different animal species and breeds are very small [47,54], differences in the mineral composition of cheese are mainly due to technological processes.

In the case of sheep cheeses, a high zinc content would be expected due to the high zinc amount of sheep milk. According to Raynal-Ljutovac et al. [55], in fact, sheep milk is richer in zinc than cow milk and goat milk, but these results are not often confirmed by other literature studies [56]. However, the sheep cheeses analysed in this study, most of which are Pecorino-type cheeses, show low zinc levels (both on a wet and dry basis). This could be due to the typical technological process of Pecorino-type cheese: in this cheesemaking after coagulation, the curd is broken into the size of a rice grain. Breaking the curd into very small granules could promote the consequent release of zinc into the whey. Furthermore, the cheese salting, which in the case of Pecorino is of greater extent compared to other cheeses [32,33,57–59], has been revealed to cause a decrease in the content of some elements, including zinc, after curdling [45]. Even in the case of Fiore Sardo PDO [60], another sheep cheese resulted with a low zinc content, the breaking of the curd is very strong and lasts for about 3 min, so as to reduce it into granules similar to a millet grain.

Concerning cheeses from goat milk, only one type of this sample was available for this study, namely Robiola di Roccaverano. Besides the fact that goat milk does not have a high zinc content, the lowest zinc amount observed in this cheese certainly reflects its manufacturing process (acid coagulation), which involves an acid coagulation before adding the rennet, resulting in a cheese characterized by high moisture content and a fragile curd [61]. Because of the acidification, the stability of casein micelles decreases, their charge becomes lower [62] with the loss of calcium and other minerals, such as zinc, in the whey. This is due to the tendency of zinc to form stable complexes with peptides and proteins [63] and it has been shown that even more than 90% of the zinc in sheep milk and about 87% of the zinc in goat milk is present in the micellar fraction, mainly bound to caseins [52], with similar distribution reported for cow milk [63]. The low zinc content in Robiola di Roccaverano is comparable with the levels detected in other cheeses with acid coagulation technology such as cottage cheese: actually, some authors founded 0.48 mg/100 g of zinc in cottage cheese [30]. Similarly, Moreno-Rojas et al. [64] reported a zinc value of 0.54 mg/100 g for Los Balanchares, a cheese produced from goat milk by acid coagulation, compared to other Spanish cheeses produced through enzymatic coagulation, the zinc levels of which ranged between 1.71 and 3.84 mg/100 g. Similarly, Lucas et al. [65] observed a very low zinc content in a goat cheese from lactic coagulation (Rocamadour) compared to other cow cheeses from rennet coagulation.

Wide variations have been reported in the zinc levels of buffalo milk [66], but data reported in the literature show that it contains moderate zinc amounts (0.57 mg/100 g [67]; 0.44 mg/100 g [68]; 0.41 mg/100 g [69]; 0.63 mg/100 g [70]) compared to milk from other species. In this study, one cheese from pure buffalo milk was evaluated, i.e., Mozzarella di Bufala Campana PDO whose values are comparable with sheep and mixed cheeses (Figure 1). According to its product specification [71], the acidification of milk and curd are obtained by adding natural whey; then maturation of the curd takes place under whey over a variable time period (maximum 5 h), during which pH gradually decreases, contributing to modify the structure and rheological properties of the curd, and promoting cheese spinning. The pH decrease of milk and curd in Mozzarella leads to a decrease in mineral content [33], therefore a not very high zinc content would be expected. This is what is observed in Mozzarella di Bufala Campana PDO on wet weight but, performing analysis on a dry weight surprisingly reveals a higher zinc content compared to sheep cheeses, as well as to some cow cheeses (Taleggio and Castelmagno). Probably the starting zinc value of the buffalo milk was very high, and the curd maturation into the whey could promote the reabsorption of the zinc lost through the whey.

2.2. *Nutritional Evaluation of Zinc in PDO and Traditional Italian Cheeses*

Dairy products are food items largely consumed: in Europe the average daily consumption of cheese is about 35.3 mg/day in adults [72], with Italy being the country with the highest cheese consumption. According to the Italian National Food Consumption Survey 2005–2006 [73], in fact, the Italian daily intake of cheese is 59.6 g for adults (18–64.9 years), with "Milk, milk products and substitutes" category providing 21% of the zinc daily intake [74].

According to the Daily Reference Intake (DRI) set by the European Parliament and the Council of the European Union [28], the requirement for zinc in adults is estimated at 10 mg/day. In order to calculate the contribute of cheese consumption to the zinc DRI, it is necessary to establish the amount of the cheese servings; however, the daily serving sizes of cheeses are not well defined all over the world, and they may vary in different countries. A recent Italian position paper [75] have defined two different servings for cheeses based on the fat amount: a serving size of 50 g if the fat content of cheese is more than 25 g/100 g edible weight, and a serving size of 100 g if the fat content is less than 25 g/100 g edible weight.

Tables 4–6 report the zinc Estimated Dietary Intakes (%) for one serving (50 or 100 g) of each cheese. Based on the European DRI [28] the percentage of the zinc intake ranges from 11.4 (Taleggio) to 38.7% (Piave Vecchio) with 50 g of cow cheeses, reaching 48.7% in the case of 100 g serving of unripened Provola delle Madonie (Table 4). Concerning cheeses from sheep milk, the percentage of the zinc intake is between 10.9% (Pecorino d'Abruzzo 7 months ripening) and 18.5% (Canestrato Pugliese) (Table 5), while in cheeses from mixed milk the data range from 1.9% (Robiola di Roccaverano) to 22.7% (Bra hard type) (Table 6). With 100 g of Ricotta Romana the percentage of the zinc intake is only 3.6%, while with a serving (100 g) of Mozzarella di Bufala Campana PDO this intake is 21.4%.

These results show that the Estimated Dietary Intake calculated for zinc in the cheese samples provides a significant contribution to the DRI set by EU [28] in most of the samples analyzed, corroborating the statement that cheeses are a good source of this essential element. Indeed, among 57 cheeses evaluated, two samples (Robiola di Roccaverano and Ricotta Romana) contribute to less than 4% of the Daily Reference Intake for zinc, 27 samples provide 10–20% of the zinc DRI, 23 cheeses ensure an average percentage of zinc intake ranging from 20 to 30%, while 5 samples even exceed 30%, with Provola delle Madonie reaching the highest value (almost 50%). These results are much higher than those reported by Gambelli et al. [30], who observed a zinc daily intake by Italian cheese consumption of about 10%. Similarly, Barone et al. [37] estimated a zinc intake ranging from 3.09% to 6.50% through consumption of some Traditional Italian cheeses. The authors [37], however, referred to a general cheese serving of 65 g/person/day and to the Recommended Dietary Allowances (RDAs) for zinc set by the Food and Nutrition Board of the Institute of Medicine [76], that are 11 mg/day for men and 8 mg/day for women. Additionally, Bilandžić et al. [77] referred to the same RDAs in evaluating the contribution of cheeses from Croatia to the zinc RDA: the authors observed very high values of EDI (%), with data ranging from 4.9% (in male through fresh cheese consumption) to 22.8% (in female through hard fat cheese consumption). However, the authors [77] referred to a cheese serving of 200 g/day per adult, so their EDIs values can be considered much lower than those reported in this work.

It must be highlighted, however, that the Daily Reference Intake for zinc set by EU [28] has some limitations: it does not consider that worldwide Zn recommendations are greater in men than in women because of the differences in protein requirement, and it does not consider the difference in diet composition and in particular the inhibitory effect of dietary fibre, phytate, or polyphenols [78] on the zinc bioavailability and absorption. Among these parameters, phytic acid is one of the most important because it chelates cations, forming insoluble complexes that cannot be digested or absorbed by humans, thus leading to a reduction in the zinc absorption. There is a wide variation in phytate intake among diets or countries: a mixed diet can contribute for about 300 to 800 mg/day of phytate, up to 1400 mg/day for diets with a high proportion of unrefined cereal grain products and legumes, while in vegetarian diets the phytate intake could reach 1600–2500 mg/day [13].

For this reason, based on the available data from zinc absorption studies and on the molar ratios phytate/zinc, WHO/FAO [14] has classified diets into three categories according to the potential bioavailability of their zinc, namely high, moderate, and low zinc availability. Consequently, three different Recommended Nutrient Intakes (RNIs) for dietary zinc (mg/day) in adults (males 19–65 years 65 kg body weight, and females 19–65 years 55 kg body weight) were set according to the three diets differing in zinc bioavailability.

However, more recently, EFSA panel has suggested different zinc Population Reference Intakes (PRIs) for adult (18–79 years) men (79.4 kg body weight) and women (68.1 kg body weight) based on four phytate intake levels (300, 600, 900, and 1200 mg/day); these PRIs range from 7.5 to 12.7 mg/day of zinc for women, and from 9.4 to 16.3 mg/day for men [13].

Table 4. Estimated Dietary Intake (%) of zinc in adults calculated according to the reference values of EU [28] and EFSA [13], with a serving (50 g or 100 g) of cheese from cow milk.

Cheeses	Serving	Adults	Male				Female			
		° DRI 10 mg/day	300 mg phytate/day § PRI 9.4 mg/day	600 mg phytate/day PRI 11.7 mg/day	900 mg phytate/day PRI 14.0 mg/day	1200 mg phytate/day PRI 16.3 mg/day	300 mg phytate/day PRI 7.5 mg/day	600 mg phytate/day PRI 9.3 mg/day	900 mg phytate/day PRI 11.0 mg/day	1200 mg phytate/day PRI 12.7 mg/day
PDO	(g)	Estimated Dietary Intake (%)								
Asiago	50	23.0	24.5	19.7	16.4	14.1	30.7	24.8	20.9	18.1
Asiago Pressato	50	20.4	21.7	17.4	14.6	12.5	27.2	21.9	18.5	16.1
Bitto	50	22.4	23.8	19.2	16.0	13.8	29.9	24.1	20.4	17.6
Caciocavallo Silano	50	22.4	23.8	19.1	16.0	13.7	29.9	24.1	20.4	17.6
Casera	50	24.6	26.2	21.1	17.6	15.1	32.8	26.5	22.4	19.4
Fontina	50	19.8	21.0	16.9	14.1	12.1	26.3	21.2	18.0	15.6
Formai de Mut	50	23.6	25.1	20.2	16.8	14.5	31.5	25.4	21.4	18.6
Gorgonzola	50	14.7	15.6	12.5	10.5	9.0	19.6	15.8	13.3	11.6
Grana Padano (>16 m)	50	27.4	29.1	23.4	19.6	16.8	36.5	29.5	24.9	21.6
Montasio	50	23.2	24.6	19.8	16.5	14.2	30.9	24.9	21.1	18.2
Monteveronese	50	20.3	21.6	17.4	14.5	12.5	27.1	21.9	18.5	16.0
Monteveronese d'allevo	50	21.5	22.8	18.3	15.3	13.2	28.6	23.1	19.5	16.9
Nostrano Valtrompia (*Alpine pasture*)	50	23.3	24.8	19.9	16.6	14.3	31.0	25.0	21.2	18.3
Nostrano Valtrompia (*indoor*)	50	25.1	26.7	21.4	17.9	15.4	33.4	26.9	22.8	19.7
Parmigiano Reggiano (>18 m)	50	26.4	28.1	22.6	18.9	16.2	35.2	28.4	24.0	20.8
Piave Vecchio	50	38.7	41.2	33.1	27.7	23.8	51.6	41.7	35.2	30.5
Provolone Valpadana	50	20.9	22.2	17.8	14.9	12.8	27.8	22.4	19.0	16.4
Puzzone Di Moena	50	21.1	22.4	18.0	15.1	12.9	28.1	22.7	19.2	16.6
Quartirolo Lombardo	50	17.1	18.2	14.6	12.2	10.5	22.8	18.4	15.5	13.4
Ragusano	50	24.2	25.8	20.7	17.3	14.9	32.3	26.1	22.0	19.1
Spressa delle Giudicarie	50	29.2	31.0	24.9	20.8	17.9	38.9	31.4	26.5	23.0
Stelvio	50	20.3	21.6	17.3	14.5	12.4	27.0	21.8	18.4	16.0
Taleggio	50	11.4	12.1	9.7	8.1	7.0	15.2	12.2	10.3	9.0
Toma Piemontese	50	19.8	21.0	16.9	14.1	12.1	26.4	21.3	18.0	15.6
TRADITIONAL										
Bagoss (*Alpine pasture)*	50	31.1	33.1	26.6	22.2	19.1	41.5	33.5	28.3	24.5
Bagoss (*indoor)*	50	29.5	31.4	25.2	21.1	18.1	39.3	31.7	26.8	23.2
Caciotta di Raiano	50	17.8	18.9	15.2	12.7	10.9	23.7	19.1	16.2	14.0
Formaggella della Val Trompia (*Alpine pasture)*	50	22.3	23.7	19.0	15.9	13.7	29.7	23.9	20.2	17.5
Formaggella della Val Trompia (*indoor)*	50	16.9	18.0	14.5	12.1	10.4	22.6	18.2	15.4	13.3
Giuncata Vaccina abruzzese	100	18.3	19.4	15.6	13.1	11.2	24.4	19.6	16.6	14.4
Provola delle Madonie unripened	100	48.7	51.8	41.6	34.8	29.8	64.9	52.3	44.2	38.3
Provola delle Madonie 4 m	50	31.3	33.3	26.7	22.3	19.2	41.7	33.6	28.4	24.6
Provola delle Madonie 16 m	50	35.9	38.2	30.7	25.6	22.0	47.8	38.6	32.6	28.3
Provola delle Madonie smoked	50	23.9	25.5	20.5	17.1	14.7	31.9	25.7	21.8	18.8

° DRI = Daily Reference Intake set by the EU Regulation [28] for zinc in adults. § PRI = Population Reference Intakes set by EFSA [13] for zinc in men and women at different phytate levels.

Table 5. Estimated Dietary Intake (%) of zinc in adults calculated according to the reference values of EU [28] and EFSA [13], with a serving (50 g or 100 g) of cheese from sheep milk.

Cheeses	Serving	Adults	Male				Female			
		° DRI 10 mg/day	300 mg phytate/day § PRI 9.4 mg/day	600 mg phytate/day PRI 11.7 mg/day	900 mg phytate/day PRI 14.0 mg/day	1200 mg phytate/day PRI 16.3 mg/day	300 mg phytate/day PRI 7.5 mg/day	600 mg phytate/day PRI 9.3 mg/day	900 mg phytate/day PRI 11.0 mg/day	1200 mg phytate/day PRI 12.7 mg/day
PDO	(g)	Estimated Dietary Intake (%)								
Canestrato Pugliese	50	18.5	19.6	15.8	13.2	11.3	24.6	19.8	16.8	14.5
Fiore Sardo	50	16.3	17.3	13.9	11.6	10.0	21.7	17.5	14.8	12.3
Formaggio di Fossa di Sogliano	50	13.6	14.5	11.6	9.7	8.3	18.1	14.6	12.4	10.7
Pecorino Romano	50	18.3	19.4	15.6	13.1	11.2	24.4	19.7	16.6	14.4
Pecorino Sardo	50	18.0	19.2	15.4	12.9	11.1	24.0	19.4	16.4	14.2
Pecorino Toscano	50	14.1	15.0	12.0	10.0	8.6	18.8	15.1	12.8	11.1
TRADITIONAL										
Pecorino d'Abruzzo Primo Sale	50	14.5	15.4	12.4	10.3	8.9	19.3	15.5	13.1	11.4
Pecorino d'Abruzzo 2 m	50	15.5	16.5	13.2	11.1	9.5	20.7	16.7	14.1	12.2
Pecorino d'Abruzzo 3 m	50	14.2	15.2	12.2	10.2	8.7	19.0	15.3	13.0	11.2
Pecorino d'Abruzzo 4 m	50	13.9	14.8	11.9	10.0	8.6	18.6	15.0	12.7	11.0
Pecorino d'Abruzzo 6 m	50	17.8	18.9	15.2	12.7	10.9	23.7	19.1	16.2	14.0
Pecorino d'Abruzzo 7 m	50	10.9	11.6	9.3	7.8	6.7	14.5	11.7	9.9	8.[illegible]
Pecorino d'Abruzzo 18 m	50	17.5	18.6	14.9	12.5	10.7	23.3	18.8	15.9	13.8
Giuncatella abruzzese	100	13.4	14.2	11.4	9.5	8.2	17.8	14.4	12.1	10.5

° DRI = Daily Reference Intake set by the EU Regulation [28] for zinc in adults. § PRI = Population Reference Intakes set by EFSA [13] for zinc in men and women at different phytate levels.

Table 6. Estimated Dietary Intake (%) of zinc in adults calculated according to the reference values of EU [28] and EFSA [13], with a serving (50 g or 100 g) of cheese from buffalo milk, from mixed milk, and from sheep whey.

Cheeses	Serving	Adults	Male				Female			
		° DRI 10 mg/day	300 mg phytate/day § PRI 9.4 mg/day	600 mg phytate/day PRI 11.7 mg/day	900 mg phytate/day PRI 14.0 mg/day	1200 mg phytate/day PRI 16.3 mg/day	300 mg phytate/day PRI 7.5 mg/day	600 mg phytate/day PRI 9.3 mg/day	900 mg phytate/day PRI 11.0 mg/day	1200 mg phytate/day PRI 12.7 mg/day
PDO	(g)	Estimated Dietary Intake (%)								
Mozzarella di Bufala Campana	100	21.4	22.8	18.3	15.3	13.1	28.6	23.0	19.5	16.9
Murazzano	50	10.3	11.0	8.8	7.4	6.3	13.8	11.1	9.4	8.1
Castelmagno	50	13.2	14.0	11.3	9.4	8.1	17.6	14.2	12.0	10.4
Bra Tenero (*soft type*)	50	18.5	19.7	15.8	13.2	11.3	24.6	19.9	16.8	14.5
Bra Duro (*hard type*)	50	22.7	24.2	19.4	16.2	13.9	30.3	24.4	20.6	17.9
Raschera	50	19.5	20.7	16.6	13.9	11.9	26.0	20.9	17.7	15.3
Caciotta d'Urbino	50	16.0	17.0	13.6	11.4	9.8	21.3	17.2	14.5	12.6
Robiola di Roccaverano	50	1.9	2.1	1.7	1.4	1.2	2.6	2.1	1.8	1.5
Ricotta Romana	100	3.6	3.8	3.1	2.6	2.2	4.8	3.9	3.3	2.3

° DRI = Daily Reference Intake set by the EU Regulation [28] for zinc in adults. § PRI = Population Reference Intakes set by EFSA [13] for zinc in men and women at different phytate levels.

According to these PRIs set by EFSA, Tables 4–6 also report the % Estimated Dietary Intakes (EDIs) for zinc in cheeses based on the four different phytate intake levels. Comparing the EDIs calculated according to the EU regulation [28] or according to the EFSA recommendation [13], it is worth observing that the zinc dietary intakes for a 600 mg/day phytate content are quite similar to the EFSA recommendations, ranging, in all studied cheeses, from 1.7 to 41.6% in male and from 2.1 to 52.3% in female.

It is well known that there are discrepancies among the worldwide recommended nutrient values: as an example, low-income countries follow the WHO recommendations, while the USA and Canada set different recommendations [79,80], and Australia and New Zealand follow their own recommendations [81]. These differences are probably due to different lifestyles and/or types of diet. The new levels proposed by EFSA could be a problem, for example, in a low-income country, where there is a high consume of unrefined cereal, while they could be easier to apply in European countries.

Since there are a lot of public health implications due to zinc deficiency, IZiNCG [2] suggested three intervention strategies: supplementation, fortification, and dietary diversification/modification. However, it is worth highlighting that, because of the coexistence of several micronutrient deficiencies, all the intervention programs should be linked each other in order to obtain the maximum effect and an efficient use of resources [2].

As reported by Freeland-Graves et al. [78], an effort could be made to harmonize dietary standards, and the enhancement of micronutrient bioavailability could be improved by the adoption of food preparation or food processing techniques.

Finally, the Estimated Daily Intake calculated in adults for each cheese was also compared to toxicological values, which in the case of zinc is the Provisional Maximum Tolerable Daily Intake (PMTDI) set by FAO/WHO [26]: all cheese samples were lower than the above mentioned PMTDI (1.00 mg/kg body weight), suggesting that, in terms of food security, the consumption of the evaluated cheeses does not cause any health hazards to consumer.

3. Materials and Methods

3.1. Samples

A total of 57 PDO and Traditional Italian cheeses were purchased from different local groceries, supermarkets, or local farms in Italy. In particular, this study evaluated cheese samples from milk of different animal species, belonging to 47 labels (38 PDO and 9 Traditional), and classified as below:

- 23 PDO and 7 Traditional cow milk cheeses
- 6 PDO and 2 Traditional sheep milk cheeses
- 1 Mozzarella PDO cheese made with buffalo milk
- 7 PDO cheeses made with milk of different animal species (various percentages of cow, sheep, and goat milk according to their production specifications)
- 1 PDO Ricotta a dairy product made with a mixture of sheep whey and milk.

For each cheese three different wheels were collected.

All dairy products were sampled and grated according to the reference method [82] and analysed in triplicate.

3.2. Chemicals

Ultrapure water was prepared by an ion exchange system to >18 mΩ resistivity (ElgaPurelab Ultra, Veolia, UK). All reagents were of the highest available purity level and the zinc standard was obtained from Merck (Darmstadt, Germany).

3.3. Zinc Analysis and Equipment

Zinc was determined according to a modification of the AOAC method [83]. The cheese samples were analysed for their zinc content after ashing: 1 g of sample were weighed into platinum crucibles, dried 1 h in 100 °C forced air oven and ashed into the furnace at 525 °C for 16 h; ashes were dissolved in 1 mL HNO_3 65% and then filled to a final volume of 50 mL with HNO_3 1% (*v/v*).

Zinc was determined using an Atomic Absorption Spectrometer AAnalyst 300 (Perkin-Elmer, Norwalk, CT, USA). A standard stock solution of zinc was used (1000 mg/L, zinc standard solution traceable to SRM from NIST). Individual working standard solutions were prepared at three different levels by dilution (0.25, 0.50, and 1.00 µg/mL) in HNO_3 1% *v/v*. The study involved a Standard Reference Material (SRM 1846 Infant formula) certified for zinc concentration.

3.4. Estimated Dietary Intake

The Estimated Dietary Intake (EDI) was considered as the percentage contribution of the zinc intake through each cheese consumption to the daily reference intakes set by different international authorities [28,72] for adults. It was calculated using the following Equation (1):

$$\text{EDI } (\%) = (\text{C} \times \text{IR}) \times 100/\text{DRI (or PRI)} \quad (1)$$

where C represents the zinc concentration in cheese (mean value expressed as mg/g), IR is the ingestion rate (g/day) of cheese, equivalent to a serving of 50 g/person/day or 100 g/person/day [75], DRI is the Daily Reference Intake set by the European Parliament and the Council of the European Union for zinc in adults [28], and PRI is the Population Reference Intakes set by the European Food Safety Authority (EFSA) for zinc in men and women at different phytate levels [13].

3.5. Statistical Analysis

One-way analysis of variance (ANOVA) test, coupled with the Tukey's post hoc test, was performed to evaluate significant differences ($p < 0.05$) among the mean values. Statistical analysis was performed using XL-STAT Base 18.06 (Addinsoft 1995–2017).

4. Conclusions

Even if in trace concentrations, zinc represents an essential element for the human body, but consumption in levels exceeding recommended concentrations could result in adverse consequences. Therefore, its intake through the diet must be carefully monitored and at the same time guaranteed. The analysis of different Italian cheeses (both PDO and Traditional) evaluated in this study showed a great variability in their zinc content, which values ranged from 0.36 to 7.75 mg/100 g cheese. Further studies will be carried out in order to confirm the obtained results. Differences in the zinc levels of cheeses may be ascribed to a several number of factors: milk origin (cow, sheep, goat, and buffalo), animal feeds quality, and characteristics of the milk, which are in turn influenced by different variables (i.e., breed, physiological condition of the animal, stage of lactation, etc.). Cheese-making processes are indeed the mainly responsible for the observed differences in the zinc content of the various cheeses, but contamination phenomena (from the process equipment, packaging, and/or storage) could also play an important role.

The cheese analysed in this work provided a great contribution to the zinc reference intakes: most of the samples (about 50%) provide more than 20% of the European Daily Reference Intake for zinc (10 mg/day) with one serving of cheese. Furthermore, the consumption of the cheese evaluated in the study do not represent a potential public health risk, since no cheeses serving exceeded the Provisional Maximum Tolerable Daily Intake set for zinc (1.00 mg/kg body weight). Therefore, the consumption of cheese as a valuable source of zinc was once again confirmed.

Author Contributions: Conceptualization, P.M.; investigation, P.M. and M.G.D.C.; methodology, P.M. and M.G.D.C.; formal analysis, P.M.; writing—original draft preparation, P.M. and M.R.; writing—review and editing, P.M. and M.R. All authors have read and agreed to the published version of the manuscript.

Funding: This research received no external funding.

Institutional Review Board Statement: Not applicable.

Informed Consent Statement: Not applicable.

Data Availability Statement: The data presented in this study are available on request from the corresponding author.

Acknowledgments: This research was partially supported by the Ministry of Agricultural, Food and Forestry Policies—BIOVITA and TERRAVITA Projects.

Conflicts of Interest: The authors declare no conflict of interest.

Sample Availability: Samples of the compounds are available from the authors.

References

1. King, J.C.; Brown, K.H.; Gibson, R.S.; Krebs, N.F.; Lowe, N.M.; Siekmann, J.H.; Raiten, D.J. Biomarkers of Nutrition for Development (BOND)-Zinc Review. *J. Nutr.* **2016**, *146*, 858S–885S. [CrossRef] [PubMed]
2. IZiNCG. Assessment of the Risk of Zinc Deficiency in Populations and Options for Its Control. In *International Zinc Nutrition Consultative Group (IZiNCG) Technical Document #1*; Hotz, C., Brown, K.H., Eds.; Food and Nutrition Bulletin: Tokio, Japan, 2004; Volume 25, (Suppl 2).
3. Maares, M.; Haase, H. A Guide to Human Zinc Absorption: General Overview and Recent Advances of In Vitro Intestinal Models. *Nutrients* **2020**, *12*, 762. [CrossRef] [PubMed]
4. Swinkels, J.W.G.M.; Kornegay, E.T.; Verstegen, M.W.A. Biology of zinc and biological value of dietary organic zinc complexes and chelates. *Nutr. Res. Rev.* **1994**, *7*, 129–149. [CrossRef] [PubMed]
5. Vallee, B.L.; Falchuk, K.H. The Biochemical Basis of Zinc Physiology. *Physiol. Rev.* **1993**, *73*, 79–118. [CrossRef] [PubMed]
6. Hotz, C.; Brown, K. Overview of Zinc Nutrition. *Food Nutr. Bull.* **2004**, *25*, S99–S129.
7. Roohani, N.; Hurrell, R.; Kelishadi, R.; Schulin, R. Zinc and its importance for human health: An integrative review. *J. Res. Med. Sci.* **2013**, *18*, 144–157.
8. Brown, K.H.; Wuehler, S.E.; Peerson, J.M. The Importance of Zinc in Human Nutrition and Estimation of the Global Prevalence of Zinc Deficiency. *Food Nutr. Bull.* **2001**, *22*, 113–125. [CrossRef]
9. de Benoist, B.; Darnton-Hill, I.; Davidsson, L.; Fontaine, O.; Hotz, C. Conclusions of the Joint WHO/UNICEF/IAEA/IZiNCG Interagency Meeting on Zinc Status Indicators. *Food Nutr. Bull.* **2007**, *28*, S480–S484. [CrossRef]
10. Brnić, M.; Hurrell, R.F.; Songré-Ouattara, L.T.; Diawara, B.; Kalmogho-Zan, A.; Tapsoba, C.; Zeder, C.; Wegmüller, R. Effect of phytase on zinc absorption from a millet-based porridge fed to young Burkinabe children. *Eur. J. Clin. Nutr.* **2017**, *71*, 137–141. [CrossRef]
11. Sanches, V.L.; Peixoto, R.R.A.; Cadore, S. Phosphorus and zinc are less bioaccessible in soy-based beverages in comparison to bovine milk. *J. Funct. Foods* **2020**, *65*, 103728. [CrossRef]
12. Davis, N.T.; Olpin, S.E. Studies on the phytate: Zinc molar contents in diets as a determinant of Zn availability to young rats. *Br. J. Nutr.* **1979**, *41*, 59–603. [CrossRef]
13. EFSA Panel on Dietetic Products Nutrition and Allergies. Scientific Opinion on Dietary Reference Values for zinc. *EFSA J.* **2014**, *12*, 3844. [CrossRef]
14. FAO/WHO. *Vitamin and Mineral Requirements in Human Nutrition*, 2nd ed.; World Health Organization: Geneva, Switzerland, 2004.
15. Foster, M.; Samman, S. Vegetarian Diets Across the Lifecycle: Impact on Zinc Intake and Status. In *Advances in Food and Nutrition Research*; Jeyakumar, H., Ed.; Academic Press: Cambridge, MA, USA, 2015; Volume 74, pp. 93–131.
16. Lonnerdal, B. Dietary Factors Influencing Zinc Absorption. *J. Nutr.* **2000**, *130*, 1378S–1383S. [CrossRef] [PubMed]
17. Raesa, K.; Knockaerta, D.; Struijsc, K.; Van Camp, J. Role of processing on bioaccessibility of minerals: Influence of localization of minerals and antinutritional factors in the plant. *Trends Food Sci. Technol.* **2014**, *37*, 32–41. [CrossRef]
18. Gibson, R.S.; Raboy, V.; King, J.C. Implications of phytate in plant-based foods for iron and zinc bioavailability, setting dietary requirements, and formulating programs and policies. *Nutr. Rev.* **2018**, *76*, 793–804. [CrossRef]
19. Sandström, B. Dietary Pattern and Zinc Supply. In *Zinc in Human Biology*; Mills, C.F., Ed.; Springer: London, UK, 1989; pp. 351–363. [CrossRef]
20. Sandstrom, B. Micronutrient interactions: Effects on absorption and bioavailability. *Br. J. Nutr.* **2001**, *85*, S181–S185. [CrossRef]
21. Miquel, E.; Farré, R. Effects and future trends of casein phosphopeptides on zinc bioavailability. *Trends Food Sci. Technol.* **2007**, *18*, 139–143. [CrossRef]
22. Bouhallab, S.; Bouglé, D. Biopeptides of milk: Caseinophosphopeptides and mineral bioavailability. *Reprod. Nutr. Dev.* **2004**, *44*, 493–498. [CrossRef]
23. Clare, D.A.; Swaisgood, H.E. Bioactive Milk Peptides: A Prospectus. *J. Dairy Sci.* **2000**, *83*, 1187–1195. [CrossRef]
24. Mendil, D. Mineral and trace metal levels in some cheese collected from Turkey. *Food Chem.* **2006**, *96*, 532–537. [CrossRef]
25. Tripathi, R.M.; Raghunath, R.; Sastry, V.N.; Krishnamoorthy, T.M. Daily intake of heavy metals by infants through milk and milk products. *Sci. Total Environ.* **1999**, *227*, 229–235. [CrossRef]

26. Joint FAO/WHO Expert Committee on Food Additives. *Evaluation of Certain Food Additives and Contaminants. Seventy-Seventh Report of the Joint FAO/WHO Expert Committee on Food Additives*; World Health Organization: Geneva, Switzerland, 2013; Volume 77.
27. GU. Gazzetta Ufficiale della Repubblica Italiana-Serie Generale, n.240 D.M. n. 350 Regolamento Recante Norme per L'individuazione dei Prodotti Tradizionali 1999. Available online: https://www.gazzettaufficiale.it/eli/id/1999/10/12/099G0423/sg (accessed on 23 September 2021).
28. EU. Regulation (EU) No 1169/2011 of the European Parliament and of the Council of 25 October 2011 on the Provision of Food Information to Consumers. Available online: https://eur-lex.europa.eu/legal-content/EN/ALL/?uri=CELEX%3A32011R1169 (accessed on 23 September 2021).
29. Manzi, P.; Mattera, M.; Di Costanzo, M.G.; Nicoli, S. La provola delle Madonie: Un formaggio della tradizione italiana. Caratterizzazione e valutazione chimico-nutrizionale. In *Proceedings of the 11° CISETA, Milan, Italy*; VELP Scientifica: Milan, Italy, 2013; pp. 348–352.
30. Gambelli, L.; Belloni, P.; Ingrao, G.; Pizzoferrato, L.; Santaroni, G. Minerals and trace elements in some Italian dairy products. *J. Food Compos. Anal.* **1999**, *12*, 27–35. [CrossRef]
31. Kedzierska-Matysek, M.; Barlowska, J.; Wolanciuk, A.; Litwinczuk, Z. Physicochemical, mechanical and sensory properties of long-ripened Polish and Italian cheeses and their content of selected minerals. *J. Elem.* **2018**, *23*. [CrossRef]
32. Bontempo, L.; Larcher, R.; Camin, F.; Hölzl, S.; Rossmann, A.; Horn, P.; Nicolini, G. Elemental and isotopic characterisation of typical Italian alpine cheeses. *Int. Dairy J.* **2011**, *21*, 441–446. [CrossRef]
33. Manuelian, C.L.; Currò, S.; Penasa, M.; Cassandro, M.; De Marchi, M. Characterization of major and trace minerals, fatty acid composition, and cholesterol content of Protected Designation of Origin cheeses. *J. Dairy Sci.* **2017**, *100*, 3384–3395. [CrossRef]
34. Coni, E.; Bocca, A.; Ianni, D.; Caroli, S. Preliminary evaluation of the factors influencing the trace element content of milk and dairy products. *Food Chem.* **1995**, *52*, 123–130. [CrossRef]
35. Coni, E.; Caroli, S.; Ianni, D.; Bocca, A. A methodological approach to the assessment of trace elements in milk and dairy products. *Food Chem.* **1994**, *50*, 203–210. [CrossRef]
36. Ianni, A.; Iannaccone, M.; Martino, C.; Innosa, D.; Grotta, L.; Bennato, F.; Martino, G. Zinc supplementation of dairy cows: Effects on chemical composition, nutritional quality and volatile profile of Giuncata cheese. *Int. Dairy J.* **2019**, *94*, 65–71. [CrossRef]
37. Barone, G.; Dambrosio, A.; Storelli, A.; Busco, A.; Ioanna, F.; Quaglia, N.C.; Giacominelli-Stuffler, R.; Storelli, M.M. Traditional Italian cheeses: Trace element levels and estimation of dietary intake. *J. Food Compos. Anal.* **2018**, *66*, 205–211. [CrossRef]
38. Ianni, A.; Martino, C.; Innosa, D.; Bennato, F.; Grotta, L.; Martino, G. Zinc supplementation of lactating dairy cows: Effects on chemical-nutritional quality and volatile profile of Caciocavallo cheese. *Asian Australas. J. Anim. Sci.* **2020**, *33*, 825–835. [CrossRef] [PubMed]
39. Coppa, M.; Chassaing, C.; Sibra, C.; Cornu, A.; Verbič, J.; Golecký, J.; Engel, E.; Ratel, J.; Boudon, A.; Ferlay, A.; et al. Forage system is the key driver of mountain milk specificity. *J. Dairy Sci.* **2019**, *102*, 10483–10499. [CrossRef] [PubMed]
40. Garmo, T.H.; FrâŠ˘slie, A.; HâŠ˘ie, R. Levels of Copper, Molybdenum, Sulphur, Zinc, Selenium, Iron and Manganese in Native Pasture Plants from a Mountain Area in Southern Norway. *Acta Agric. Scand.* **1986**, *36*, 147–161. [CrossRef]
41. Baranova, A.; Oldeland, J.; Wang, S.-l.; Schickhoff, U. Grazing impact on forage quality and macronutrient content of rangelands in Qilian Mountains, NW China. *J. Mt. Sci.* **2018**, *16*, 43–53. [CrossRef]
42. Noulas, C.; Tziouvalekas, M.; Karyotis, T. Zinc in soils, water and food crops. *J. Trace Elem. Med. Biol.* **2018**, *49*, 252–260. [CrossRef]
43. Martino, C.; Ianni, A.; Grotta, L.; Pomilio, F.; Martino, G. Influence of Zinc Feeding on Nutritional Quality, Oxidative Stability and Volatile Profile of Fresh and Ripened Ewes' Milk Cheese. *Foods* **2019**, *8*, 656. [CrossRef]
44. Xu, N.N.; Yang, D.T.; Miao, C.; Valencak, T.G.; Liu, J.X.; Ren, D.X. Organic zinc supplementation in early-lactation dairy cows and its effects on zinc content and distribution in milk and cheese. *JDS Commun.* **2021**, *2*, 110–113. [CrossRef]
45. Coni, E.; Bocca, A.; Coppolelli, P.; Caroli, S.; Cavallucci, C.; Trabalza Marinucci, M. Minor and trace element content in sheep and goat milk and dairy products. *Food Chem.* **1996**, *57*, 253–260. [CrossRef]
46. Moreno-Rojas, R.; Amaro-Lopez, M.; Zurera-Cosano, G. Copper, iron and zinc variations in Manchego-type cheese during the traditional cheese-making process. *Food Chem.* **1994**, *49*, 67–72. [CrossRef]
47. Muehlhoff, E.; Bennett, A.; McMahon, D. *Milk and Dairy Products in Human Nutrition*; Food and Agriculture Organization of the United Nations (FAO): Rome, Italy, 2013.
48. Pertoldi Marietta, G.; Gabrielli Favretto, L. Effects of processing on the occurrence of heavy metals in milk products. *Proc. Euro Food Chem II* **1983**, 305–314.
49. Lucas, A.; Agabriel, C.; Martin, B.; Ferlay, A.; Verdier-Metz, I.; Coulon, J.B.; Rock, E. Relationships between the conditions of cow's milk production and the contents of components of nutritional interest in raw milk farmhouse cheese. *Le Lait* **2006**, *86*, 177–202. [CrossRef]
50. Samaržija, D.; Antunac, N.; Pecina, M.; Mioc, B.; Havranek, D.; Pavlovic, I. Mineral value of Croatian artisanal hard sheep cheeses in terms of geographical indication. *Milchwissenschaft* **2005**, *60*, 158–161.
51. Bontempo, L.; Barbero, A.; Bertoldi, D.; Camin, F.; Larcher, R.; Perini, M.; Sepulcri, A.; Zicarelli, L.; Piasentier, E. Isotopic and elemental profiles of Mediterranean buffalo milk and cheese and authentication of Mozzarella di Bufala Campana PDO: An initial exploratory study. *Food Chem.* **2019**, *285*, 316–323. [CrossRef]

52. De la Fuente, M.; Olano, A.; Juarez, M. Distribution of calcium, magnesium, phosphorus, zinc, manganese, copper and iron between the soluble and colloidal phases of ewe's and goat's milk. *Le Lait* **1997**, *77*, 515–520. [CrossRef]
53. Coni, E.; Bocca, B.; Caroli, S. Minor and trace element content of two typical Italian sheep dairy products. *J. Dairy Res.* **1999**, *66*, 589–598. [CrossRef] [PubMed]
54. Martín-Hernández, C.; Amigo, L.; Martín-Alvarez, P.J.; Juárez, M. Differentiation of milks and cheeses according to species based on the mineral content. *Z. Für Lebensm. Unters. Und Forsch.* **1992**, *194*, 541–544. [CrossRef]
55. Raynal-Ljutovac, K.; Lagriffoul, G.; Paccard, P.; Guillet, I.; Chilliard, Y. Composition of goat and sheep milk products: An update. *Small Rumin. Res.* **2008**, *79*, 57–72. [CrossRef]
56. Nečemer, M.; Potočnik, D.; Ogrinc, N. Discrimination between Slovenian cow, goat and sheep milk and cheese according to geographical origin using a combination of elemental content and stable isotope data. *J. Food Compos. Anal.* **2016**, *52*, 16–23. [CrossRef]
57. Catillo, G.; Di Costanzo, M.G.; Manzi, P.; Pizzoferrato, L. Caratteristiche analitiche e composizione di formaggi DOP italiani al consumo. In Proceedings of the VII Congresso Nazionale di Chimica degli Alimenti, Perugia, Italy, 23–26 June 2008; pp. 1–6.
58. Mattera, M.; Durazzo, A.; Nicoli, S.; Di Costanzo, M.G.; Manzi, P. Chemical, nutritional, physical and antioxidant properties of Pecorino d'Abruzzo cheese. *Ital. J. Food Sci.* **2016**, *28*, 579–597.
59. Quattrucci, E.; Bruschi, L.; Manzi, P.; Aromolo, R.; Panfili, G. Nutritional Evaluation of Typical and Reformulated Italian Cheese. *J. Sci. Food Agric.* **1997**, *73*, 46–52. [CrossRef]
60. GU. Gazzetta Ufficiale Della Repubblica Italiana-Serie Generale, n.72. 1975, pp. 36–45. Available online: https://www.gazzettaufficiale.it/atto/serie_generale/caricaDettaglioAtto/originario?atto.dataPubblicazioneGazzetta=1975-03-14&atto.codiceRedazionale=1371Q008&elenco30giorni=false (accessed on 23 September 2021).
61. Lucey, J. *Cheese | Acid-and Acid/Heat Coagulated Cheese In Encyclopedia of Dairy Sciences*; Academic Press: San Diego, CA, USA, 2011; pp. 698–705.
62. Lucey, J.A. Acid coagulation of milk. In *Advanced Dairy Chemistry*; Springer: Berlin/Heidelberg, Germany, 2016; pp. 309–328.
63. Pomastowski, P.; Sprynskyy, M.; Buszewski, B. The study of zinc ions binding to casein. *Colloids Surf. B Biointerfaces* **2014**, *120*, 21–27. [CrossRef] [PubMed]
64. Moreno-Rojas, R.; Sánchez-Segarra, P.J.; Cámara-Martos, F.; Amaro-López, M.A. Multivariate analysis techniques as tools for categorization of Southern Spanish cheeses: Nutritional composition and mineral content. *Eur. Food Res. Technol.* **2010**, *231*, 841–851. [CrossRef]
65. Lucas, A.; Rock, E.; Chamba, J.-F.; Verdier-Metz, I.; Brachet, P.; Coulon, J.-B. Respective effects of milk composition and the cheese-making process on cheese compositional variability in components of nutritional interest. *Lait* **2006**, *86*, 21–41. [CrossRef]
66. Abd El-Salam, M.H.; El-Shibiny, S. A comprehensive review on the composition and properties of buffalo milk. *Dairy Sci. Technol.* **2011**, *91*, 663–699. [CrossRef]
67. Esposito, M.; Miedico, O.; Cavallo, S.; Pellicanò, R.; Rosato, G.; Baldi, L.; Chiaravalle, A.E. Trace elements in raw milk of buffaloes (Bubalus bubalis) from Campania, Italy. *Food Chem.* **2017**, *233*, 378–384. [CrossRef]
68. Enb, A.; Abou Donia, M.A.; Abd-Rabou, N.S.; Abou-Arab, A.A.K.; El-Senaity, M.H. Chemical Composition of Raw Milk and Heavy Metals Behavior During Processing of Milk Products. *Glob. Vet.* **2009**, *3*, 268–275.
69. Patiño, E.M.; Pochon, D.O.; Faisal, E.L.; Cedrés, J.F.; Mendez, F.I.; Stefani, C.G.; Crudeli, G. Influence of breed, year season and lactation stage on the buffalo milk mineral content. *Ital. J. Anim. Sci.* **2007**, *6*, 1046–1049. [CrossRef]
70. Singh, P.; Gupta, P.; Sharda, D. Zinc, copper and iron content in buffalo milk of Hyrana region. *Indian J. Dairy Sci.* **1981**, *34*, 394–396.
71. GU. Gazzetta Ufficiale Della Repubblica Italiana-Serie Generale, n.47. 2008. Available online: https://www.gazzettaufficiale.it/eli/gu/2008/02/25/47/sg/pdf (accessed on 23 September 2021).
72. EFSA (European Food Safety Authority). Food Consumption Database. Available online: https://www.efsa.europa.eu/en/data-report/food-consumption-data (accessed on 23 September 2021).
73. Leclercq, C.; Arcella, D.; Piccinelli, R.; Sette, S.; Le Donne, C. The Italian National Food Consumption Survey INRAN-SCAI 2005–06: Main results in terms of food consumption. *Public Health Nutr.* **2009**, *12*, 2504–2532. [CrossRef] [PubMed]
74. Sette, S.; Le Donne, C.; Piccinelli, R.; Arcella, D.; Turrini, A.; Leclercq, C.; Group, I.-S.S. The third Italian national food consumption survey, INRAN-SCAI 2005–06–part 1: Nutrient intakes in Italy. *Nutr. Metab. Cardiovasc. Dis.* **2011**, *21*, 922–932. [CrossRef] [PubMed]
75. SINU (Societa' Italiana di Nutrizione Umana). *LARN—Livelli di Assunzione di Riferimento di Nutrienti ed Energia per la Popolazione Italiana*; SICS Editore Srl: Milan, Italy, 2014.
76. Trumbo, P.; Yates, A.A.; Schlicker, S.; Poos, M. Dietary Reference Intakes: Vitamin A, Vitamin K, Arsenic, Boron, Chromium, Copper, Iodine, Iron, Manganese, Molybdenum, Nickel, Silicon, Vanadium, and Zinc. *J. Am. Diet. Assoc.* **2001**, *101*, 294–301. [CrossRef]
77. Bilandžić, N.; Sedak, M.; Đokić, M.; Božić, Đ.; Vrbić, A. Content of macro-and microelements and evaluation of the intake of different dairy products consumed in Croatia. *J Food Compos. Anal.* **2015**, *40*, 143–147. [CrossRef]
78. Freeland-Graves, J.H.; Sachdev, P.K.; Binderberger, A.Z.; Sosanya, M.E. Global diversity of dietary intakes and standards for zinc, iron, and copper. *J. Trace Elem. Med. Biol.* **2020**, *61*, 126515. [CrossRef] [PubMed]

79. Canada.ca. Dietary Reference Intakes. Available online: https://www.canada.ca/en/health-canada/services/food-nutrition/healthy-eating/dietary-reference-intakes.html (accessed on 23 September 2021).
80. U.S. Department of Health and Human Services; U.S. Department of Agriculture. 2015–2020 Dietary Guidelines for Americans. Available online: https://health.gov/our-work/food-nutrition/previous-dietary-guidelines/2015 (accessed on 23 September 2021).
81. National Health and Medical Research Council; Australian Government Department of Health and Ageing; New Zealand Ministry of Health. *Nutrient Reference Values for Australia and New Zealand*; National Health and Medical Research Council: Canberra, Australia, 2006.
82. ISO 707/IDF 50. *Milk and Milk Products—Guidance on Sampling*; International Organization for Standardization & International Dairy Federation: Geneva, Switzerland, 2008.
83. AOAC (Association of Official Analytical Chemists). *Method 991.25. Calcium, Magnesium and Phosphorus in Cheese*; Official Methods of Analysis of the Association of Official Analytical Chemists: Arlington VA, USA, 2002.

molecules

MDPI

Opinion

Vitamin D Fortification of Consumption Cow's Milk: Health, Nutritional and Technological Aspects. A Multidisciplinary Lecture of the Recent Scientific Evidence

Luisa Pellegrino [1,*], Franca Marangoni [2], Giovanna Muscogiuri [3,4], Paolo D'Incecco [1], Guillaume T. Duval [5], Cedric Annweiler [5,6] and Annamaria Colao [3,4]

1 Department of Food, Environmental and Nutritional Sciences, University of Milan, Via G. Celoria 2, 20133 Milan, Italy; paolo.dincecco@unimi.it
2 NFI—Nutrition Foundation of Italy, Viale Tunisia 38, 20124 Milan, Italy; marangoni@nutrition-foundation.it
3 Department of Clinical Medicine and Surgery, Federico II University of Naples, Via S. Pansini 5, 80131 Naples, Italy; giovanna.muscogiuri@unina.it (G.M.); colao@unina.it (A.C.)
4 UNESCO Chair for Health Education and Sustainable Development, Federico II University of Naples, Via S. Pansini 5, 80131 Naples, Italy
5 Department of Geriatric Medicine and Memory Clinic, Research Center on Autonomy and Longevity, University Hospital, 49035 Angers, France; duval.guillaume@hotmail.fr (G.T.D.); CeAnnweiler@chu-angers.fr (C.A.)
6 Department of Medical Biophysics, Robarts Research Institute, Schulich School of Medicine and Dentistry, The University of Western Ontario, London, ON N5X 4L2, Canada
* Correspondence: luisa.pellegrino@unimi.it

Abstract: Vitamin D is essential in assuring bone health at all stages of life, but its non-skeletal effects are also essential: This vitamin impacts the physiology of the immune system, skeletal muscles and adipose tissue, glucose metabolism, skin, cardiovascular and reproductive systems, neuro-cognitive functions and cell division. The incidence of vitamin D deficiency is widespread worldwide, at any age, in young and healthy subjects, as well as in pregnant women and the elderly population, due to several factors, including inadequate sunlight exposure, skin pigmentation and coverage, adiposity, lifestyle and low dietary intakes. To overcome this problem, the fortification of foods that are consumed on a daily basis, such as milk, is strongly advisable. This opinion paper aims to discuss, in a multidisciplinary way, the current evidence supporting the importance of vitamin D in health and disease and the role of milk as an optimal carrier of this vitamin, to promote adequate intakes, highlighting its unique physico-chemical characteristics linked to both fat globule membrane and casein micelle structure. Moreover, it addresses the impact of industrial processing and storage of consumption milk on the stability of these structures, thus in determining vitamin D bioavailability and the achievement of adequate intakes.

Keywords: vitamin D; bone health; vitamin D deficiency; milk fortification; consumption milk

1. Introduction

Vitamin D deficiency has become a relevant problem worldwide, at any age: In the elderly population, but also in young and healthy subjects; in fact, the endogenous synthesis of the vitamin may be inadequate even in people living in sunny regions, also due to lifestyle changes occurring in last decades [1–3]. Such deficiency, as a consequence, is one of the emerging and most widespread problems of public health, since vitamin D is the main compound responsible for the regulation and control of calcium metabolism and bone health [4,5]. In addition, accumulating evidence describe its role in the regulation of the immune system activity, and specifically in its ability to inhibit the production of inflammatory cytokines and to act in all organs of the human body, contributing to the reduction of the incidence of many common diseases [6,7].

To overcome the limited intakes of vitamin D with common diets, the fortification of foods consumed on a regular basis is frequently adopted to improve the vitamin D status at the population level [8]. Among them, cow's milk, which represents an important food for a large segment of world population [9], is characterized by high nutrient density at an affordable price [10,11]. In addition, modern technologies for milk sanitation and packaging allow consumption milk to be stored at room temperature up to 10–12 months, thus supplying this food within an increasing trading area. In this context, milk can be a suitable tool to contribute daily to the adequate intake of vitamin D in order to maintain the homeostasis in the blood, reducing the possibility of insufficiency or deficiency. Even if the natural content of vitamin D in cow's milk is low, i.e., 0.1–1 μg/L in full fat milk [12,13], the fortification of consumption milk with vitamin D has shown to be effective in improving the vitamin intake in many countries [14]. Nowadays, this practice is mandatory in some countries, while in others, is strongly recommended [14].

The aim of this work is to discuss, from a multidisciplinary point of view, the wealth of information currently available on both nutritional and health aspects of vitamin D and the role of cow's milk as a natural carrier of the vitamin D in human diet.

2. Vitamin D

Vitamin D is structurally a steroid-like molecule, with a side chain that is unsaturated in vitamin D2 (ergocalciferol, of vegetable origin) and saturated in vitamin D3 (cholecalciferol, of animal origin); it is biologically inactive and requires double hydroxylation in order to be converted to the active form, able to affect mineral metabolism and to control many other diverse physiological functions [15].

Vitamin D is mostly synthesized in the skin from 7-dehydrocholesterol by ultraviolet irradiation. It binds to an alpha globulin, known as vitamin-D binding protein (DPB), and a small amount is obtained from the diet. It is transported to the liver, where it undergoes the first hydroxylation, and then to the proximal renal tubules, where the second hydroxylation is operated by the 1αhydroxylase, thus resulting in the hormonally active form, namely 1,25-dihydroxyvitamin D (1,25 (OH)2D), which is responsible for most of the biological actions of vitamin D [16].

Probably, due to its widespread effects, 1,25(OH)2D is tightly regulated in its bioavailability and in the processes of activation and deactivation, through a series of negative and positive feedbacks that principally occur in the proximal renal tubule [17]. The negative feedback mechanisms depend on 1,25(OH)2D3 itself, phosphorus, calcium and fibroblast growth factor 23 (FGF-23), whereas the positive ones depend on the parathyroid hormone (PTH), calcium and the type 1 insulin-like growth factor (IGF-1). Low dietary calcium and phosphate result in enhanced activity of 1αhydroxylase. Elevated PTH resulting from hypocalcemia is a primary signal mediating the induction of 1,25(OH)2D synthesis in the kidney. In addition, FGF23 that promotes renal phosphate excretion is also a physiological regulator of vitamin D metabolism. By inhibiting synthesis and promoting catabolism of 1,25(OH)2D, FGF23 functions to reduce the levels of 1,25(OH)2D, which in turn decreases FGF23 expression in bone, forming a negative feedback circuit between FGF23 and the vitamin D endocrine system [18,19].

3. Vitamin D and Bone Health

Bone effects of vitamin D have been widely studied. They were actually recognized and explored long before non-bone effects. Vitamin D is involved in bone health at all stages of life, and vitamin D deficiency results in adverse effects on bone structure and global health. It is therefore particularly important to improve overall knowledge of both the prescribers and the population on the bone effects of vitamin D during life span. Vitamin D plays an essential role in the control of bone remodeling through the regulation of phospho-calcium metabolism.

3.1. Biomolecular Effects of Vitamin D on Bone

The active metabolite of vitamin D (1,25(OH)2D) stimulates the synthesis of the calcium-binding protein and controls the opening of calcium channels in the intestinal wall leading to an absorption of calcium and phosphate from the intestinal lumen to the bloodstream. These conditions are crucial for bone mineralization [20]. Vitamin D stimulates the activity of bone osteoblasts and increases the production and the plasma concentration of FGF23. Another aspect of the kidney–bone axis is highlighted here involving the active form of vitamin D. FGF23 stimulates the decrease in phosphatemia by urinary excretion and a decrease in circulating PTH. Thus, the bone remodeling activity, which results in bone loss, is reduced [21].

Conversely, in the case of vitamin D deficiency, the absorption of calcium from the digestive tract is reduced, thus reducing the amount of calcium available for bone mineralization. This decrease in serum calcium leads to an increase in PTH (i.e., secondary hyperparathoidism). The increase in serum PTH stimulates the hydroxylation of 25(OH)D to 1,25(OH)2D using the 1αhydroxylase in the proximal tubule of the kidney and promotes bone remodeling, which helps to compensate for the decrease in calcium absorption and restore serum calcium concentration [22]. The stimulation of bone remodeling by PTH during vitamin D deficiency comes at the expense of increased bone loss, which is largely irreversible, and leads to osteoporosis [23]. Exposure to prolonged vitamin D deficiency causes an increased bone turnover, which contains less mineralized osteons and more unmineralized osteoid tissue. The buildup of osteoid tissue associated with prolonged vitamin D deficiency results in osteomalacia [22].

At last, Vitamin D Receptor (VDR) gene polymorphism probably has a responsibility for differences in bone mineral density from one individual to another. Indeed, depending on the allelic variant of the VDR gene and the calcium intake, bone mineral density varies [24].

There is thus a positive linear relationship between 25(OH)D concentration and bone mineral density (BMD) within physiological limits (i.e., 25(OH)D < 100–120 ng/mL). In practice, there are two critical thresholds to fully understand this relationship (Figure 1): A threshold of "calcium homeostasis" corresponding to the stimulation of the activity of kidney 1αhydroxylase, when the concentration of 25(OH)D is above 10 ng/mL (mineralization action), and a threshold of "bone homeostasis" corresponding to the stimulation of the activity of bone 1αhydroxylase and to the increase in the expression of bone VDRs when the concentration of 25(OH)D is above 30 ng/mL (bone gain action). The intermediate concentration of 25(OH)D, around 20 ng/mL, helps prevent secondary hyperparathyroidism and therefore maintains trabecular volume.

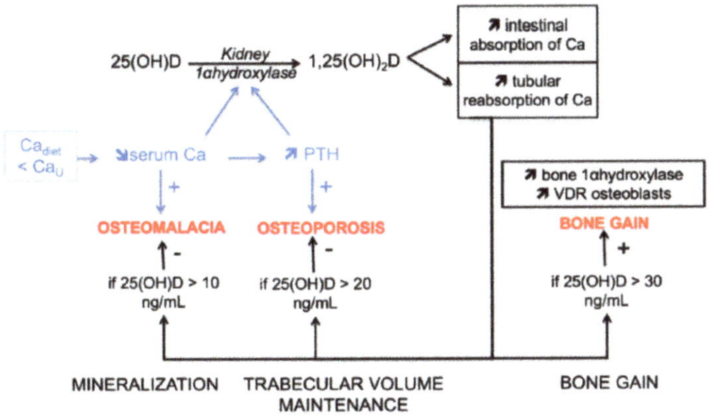

Figure 1. Bone effects of vitamin D according to the different thresholds of 25-hydroxyvitamin D concentration. 1,25(OH)2D: 1,25-dihydroxyvitamin D, 25(OH)D: 25-hydroxyvitamin D, Ca: Calcium, Cadiet: Dietary calcium intake, CaU: Urinary calcium loss, PTH: Parathyroid hormone, +: stimulation, −: Prevention.

3.2. Clinical Effects of Vitamin D Deficiency on Bone According to Age

(a) Infants and young children

The role of vitamin D in building bone in fetuses and newborns has long been known. The first cases of rickets described date from the 17th century. In fact, rickets is a disease of bone growth and mineralization resulting from vitamin D and calcium deficiency during the bone development period of infants and young children [25]. Vitamin D deficiency in this population is mainly linked to a lack of sun exposure and a low vitamin D diet. Different socio-geographic factors can lead to vitamin D deficiency in pregnant women, infants and young children. Clinical consequences of rickets are bone deformations affecting the entire skeleton but mainly affecting long bones and metaphysis growth cartilages. These abnormalities can lead to dwarfism in some extreme cases.

(b) Adults

In adults, clinical signs suggestive of a vitamin D deficiency are mainly bone discomfort or pain that is generalized or often localized in the pelvis, the lumbar region and the lower limbs, impaired physical function, muscle pain and muscle fatigue [26,27].

(c) Elderly people

The prevalence of vitamin D deficiency is very high, up to 90%, in older adults [28]. Elderly people are more at risk of vitamin D deficiency due to a frequent decrease in sun exposure, a decreased absorption of vitamin D from food and a decrease of skin vitamin D synthesis and kidney hydroxylation [29]. Vitamin D deficiency leads to secondary hyperparathyroidism, which is responsible for bone loss. This bone loss can lead to mineralization defect, osteoporosis and osteomalacia in the case of prolonged exposure. Vitamin D deficiency also exposes one to a risk of fractures due to bone fragility and an increased risk of muscle fatigability [30] and falls [31].

3.3. Bone Effects of Vitamin D Supplementation According to Age

The administration of vitamin D supplements, to prevent or control symptoms of deficiency, has a role in conditions that can be found practically at any age.

(a) Infants and young children

Vitamin D supplementation in infants and young children is mainly to prevent rickets. Before the development of synthetic vitamin D, the consumption of cod liver oil rich in vitamin D was recommended. Daily vitamin D supplementation is usually proposed in the last trimester of pregnancy for pregnant women and for newborns from birth to 18 months of age. Thereafter, more spaced treatment, often in the winter months, is proposed until the end of bone growth. This last point depends on risk factors and socio-geographic status. It is also interesting to note that exclusive breastfeeding is at greater risk of vitamin D deficiency than when using formula milk very often supplemented with vitamin D.

According to the Scientific Advisory Committee on Nutrition (SACN) and the National Institute for health and Care Excellence (NICE) Guidelines, a reference nutrient intake (RNI) of 8.5 to 10 μg (15.2 to 17.8 nanomoles) per day is recommended for all infants from birth to 1 year of age, and 10 μg (17.8 nanomoles) per day for children aged 1 to 4 years [7,26]. Nutritional intake of vitamin D refers to the consumption of foods naturally rich in vitamin D, fortified products such as growth milk and supplementation.

Higher daily doses of vitamin D or loading doses are given transitorily in the treatment of rickets. A maintenance treatment is then implemented as a follow-up to the curative treatment. Calcium supplementation may also be proposed. Orthopedic surgery is sometimes necessary, especially for severe genu valgum and genu varum causing locomotor disorders [32].

(b) Adults

In healthy middle-aged adults, reference nutrient intakes are established to ensure a sufficient vitamin D concentration for most of the population in order to avoid the harmful musculoskeletal effects of vitamin D deficiency. These recommendations of 10 micrograms (17.8 nanomoles) per day for adults are also applicable to a larger population aged 4 years and older. Reference nutrient intakes are the same for pregnant and lactating women and population groups at increased risk of vitamin D deficiency. However, these guidelines do not consider sun exposure, season, latitude and complex parameters affecting the skin synthesis of vitamin D [7,26]. In this population subgroup, vitamin D drug supplementation should not be universal but implemented only in individuals with vitamin D deficiency. Varied diet and outdoor activities should be encouraged.

(c) Elderly people

In the older population, blood testing is recommended in those at risk of or exhibiting suggestive symptoms of vitamin D deficiency or insufficiency [7]: Osteoporosis, rickets, osteomalacia, fragility fracture, osteoporosis inducing treatment or repeated falls. Other guidelines also recommend to systematically supplement all residents in nursing-homes [33].

Vitamin D supplementation in this population reduces the risks of osteomalacia, osteoporosis and fractures. Oral vitamin D supplementation between 700 and 800 UI/day is necessary to reduce the risk of hip and other non-vertebral fractures [34]. This preventive effect is dose dependent [35] and found in those with vitamin D deficiency at baseline [36]. In a meta-analysis of 11 observational studies (39,141 participants, 6278 fractures, 2367 hip fractures), each increase of 10.0 ng/mL (i.e., 25 nmol/L) in 25(OH)D concentration was associated with an adjusted relative risk (RR) for any fracture of 0.93 (95% CI, 0.89–0.96) and an adjusted RR for hip fracture of 0.80 (95% CI, 0.75–0.86) [37]. The combination of vitamin D with calcium may be more efficient than vitamin D alone since the latter meta-analysis also found that daily supplementation with both vitamin D and calcium (6 randomized clinical trials with 49,282 participants, 5449 fractures, 730 hip fractures) was associated with a 6% reduced risk of any fracture (RR, 0.94; 95% CI, 0.89–0.99) and a 16% reduced risk of hip fracture (RR, 0.84; 95% CI, 0.72–0.97) [37]. However, it is noticeable that vitamin D supplementation exhibits only a low-amplitude effect on the bone mineral density of 0–2%, which may not explain the prevention of fractures [38], the latter one being likely explained by the prevention of falls in older people [35,36].

4. Vitamin D and Non-Skeletal Effects

Although the role of vitamin D in calcium-phosphorus metabolism and in the homeostasis of bone mineral reserves is the most well-known, its effects on the homeostasis of non-skeletal tissues are equally documented. VDR are in fact ubiquitous, being widely ex-pressed in all tissues, including breast, endothelium, vascular smooth muscle, cardiomyocytes and the male urogenital system. Indeed, it has been observed that vitamin D plays a fundamental role in the physiology of the immune system, skeletal muscles and adipose tissue, glucose metabolism, skin, cardiovascular and reproductive systems, neuro-cognitive functions and cell division and sleep regulation [39–41].

The role of vitamin D on the immune system has been the subject of a breadth of literature [42] and of a scientific opinion of the European Food Safety Authority (EFSA), which approved the use of the following claim: "Vitamin D contributes to the normal function of the immune system and healthy inflammatory response" for foods which contain significant amounts of the vitamin [6].

All the cells of the immune system, as far as it is known to date, express VDR. The cell presenting antigen (APC), upon immune stimuli, are capable of producing 1,25(OH) 2D3 through the same enzyme expressed in the kidney. 1,25(OH)2D3 exerts its action on both the innate and the acquired immunity. On innate immunity, vitamin D and its metabolites stimulate the differentiation and activation of macrophages, producing defensins, such as cathelicidin and β2-defensin [43]. The main effects of vitamin D on acquired immunity are inhibitory, causing a phenotypic shift of T cells from an effector phenotype, involved in autoimmune diseases, to a regulatory and protective one.

5. Vitamin D Status: Role of the Dietary Intake

5.1. Epidemiology

Available data suggest that vitamin D deficiency is widespread worldwide, with the highest prevalence in Asia, the Middle East and Africa, as well as among immigrants from these regions living in countries at higher latitudes [44]. Even if data regarding vitamin D status were mainly from studies on selected small samples, there is consistent evidence suggesting that the risk of vitamin D deficiency would affect a larger part of general population, as shown by the ODIN project [45], funded by the European Commission. The map of vitamin D status in different countries (from 107 studies involving more than 630,000 subjects) shows that, overall, one person in eight is at risk of vitamin D deficiency, even in sunny regions [1].

The overall pooled estimate, irrespective of age group, ethnic mix and latitude registered serum 25(OH)D concentrations <12 ng/mL in 13.0% of the total European population on average in the year, and specifically in 17.7% of individuals sampled during the extended winter period (October–March) and 8.3% in those sampled in summer (April–November). In the same population, the mean prevalence rises to 40.4% if considering an alternate suggested definition of vitamin D deficiency (<20 ng/mL) [1].

This observation is confirmed by another survey by the International Osteoporosis Foundation: Circulating levels of vitamin D are, on average, below 20 ng/mL, i.e., in the range considered 'insufficient', also in populations living in the Mediterranean area [46].

Outside Europe, the risk of vitamin D deficiency has also been described in the Middle East, China, Mongolia and India, in risk groups including older persons and pregnant women. Adequate vitamin D status, defined as serum 25-hydroxyvitamin D greater than 20 ng/mL, is present in less than 50% of the world population, at least in winter [3]. Within a nationally representative sample of Australian adults aged ≥25 years (the 2011–2013 Australian Health Survey, n 5034), 20% of participants (19% of men and 21% of women) were classified as vitamin D deficient (serum 25(OH)D concentrations <20 ng/mL), with a further 43% (45% of men and 42% of women) classified as insufficient (Vitamin D: 20 to 30 ng/mL).

In the US, almost one-fifth (18.3%) of the NHANES 2011–2014 population (>1 y to >60 y) had serum 25(OH)D values categorized as at risk of inadequacy (12–20 ng/mL) and 5.0% were at risk of deficiency (<12 ng/mL) [2]. In particular, among adults aged 20–39, 7.6% were at risk of deficiency and 23.8% at risk of inadequacy; the prevalence is slightly lower among people aged 40–59 (5.7% and 18.6% at risk of deficiency and inadequacy respectively). Vitamin D deficiency has been reported to increase in the elderly, not only due to reduced skin production of vitamin D with age, but also because of age-related factors that can result in limited sun exposure, such as being more housebound [47]. In fact, some studies have shown that the risk of vitamin D deficiency increases up to 80% in institutionalized elderly [48]. A high prevalence of vitamin D deficiency has been described even among pregnant women in the Mediterranean regions (50–65% in most studies), with a certain variability between different European countries, resulting in severe skeletal and non-skeletal health events among offspring [49]. A similar prevalence has been described in the US on average, with great differences between white (13%), Hispanic (45%) and black (80%) pregnant women. As regards children and adolescents, for some years now pediatricians have been pointing out that vitamin D deficiency occurs commonly among these population groups, even in Europe. Limited data on vitamin D concentrations and vitamin D deficiency among the healthy pediatric population are available from several countries, including Denmark, England, Finland, France, Germany, Greece, Ireland, Italy, the Netherlands, Poland, Spain, Switzerland and Turkey [50]. Although some of the studies included only small numbers of children, they allow to conclude that the main causes are obesity, dark skin and inadequate sun exposure due to living in northern latitudes, excessive use of sunscreen with high SPF, staying indoors for much of the day and wearing clothes covering most of the skin [50]. More systematic information has been collected in the US where, according to NHANES 2011–2014 data,

6.6% of children aged 1–5, 12.3% of children aged 6–11 and 22.7% of adolescents (12–19 y) are at risk of inadequacy [51]. The increase in prevalence of vitamin D inadequacy (and deficiency) throughout the various age groups has been confirmed in Great Britain: According to data from the National Diet and Nutrition Survey (NDNS) (1102 samples) vitamin D circulating levels decreased progressively with age in both sexes, moving from 30 ng/mL (4–8 y) to 25.2 (9–13 y) to 16.5 (14–18 y) on average [52].

5.2. Nutritional Recommendations and Dietary Intake

The role of vitamin D intake through the diet in cases of inadequate exposure to sunlight has been emphasized by the EFSA, which has recognized an evidence-based direct association between the intake of significant amounts of the vitamin and some important health effects for the general population [4]: Both at the musculoskeletal level—by contributing to normal absorption/use of calcium and phosphorus, normal blood calcium levels, maintenance of normal bones and teeth and normal muscle function—and on the immune system, by contributing to its normal function (in infants and children too). The corresponding health claims are therefore authorized by the European Commission for products that are at least the "source" of the vitamin, i.e., containing at least 15% per 100 g or 100 mL or per consumption unit, of the nutrient reference value set at 5 µg/day [53].

The tolerable upper intake level of vitamin D has been set at 100 µg/day (4000 IU) by both EFSA in 2012 and the Institute of Medicine (IOM) in 2011 for adolescents and adults, including pregnant and lactating women [5,54]. Lower tolerable levels have been defined for infants (EFSA: 25 µg/day; IOM: 62.5 µg/day) and for children (EFSA: 50 µg/day for 1–10 years; IOM: 75 µg/day for 1–8 years).

The reference value is slightly lower than the intake of vitamin D that is considered adequate to maintain normal serum levels of 25(OH)D when skin synthesis is minimal, which is set, in Europe, at 10 µg/day for children 7–11 months of age, the same as in the United States, and to 15 µg/day for children aged 1–17 years and adults (including pregnant and lactating women), to allow the majority of the population to achieve a serum 25(OH)D concentration near or above the target of 50 nmol/L [55]. Because of lack of sun exposure and the decline with age of endogenous vitamin D synthesis, a higher reference intake, i.e., 20 µg/day, is set for people over 65 years of age, by the Institute of Medicine, the Health Council of the Netherlands, the Italian Society of Human Nutrition and the Nordic Council of Ministers [5,55]. Requirements of food-derived vitamin D may be lower with regular exposure to UVB-containing sunlight and when skin synthesis is adequate [44].

However, dietary sources of vitamin D are limited: The best available natural source is cod liver oil (90–250 µg/100g), followed by fatty fish (e.g., 6–10 µg/100 g of aquaculture salmon); lower concentrations are in lean fish, eggs, meat and dairy products [56]. Fortified foods make the highest contribution to vitamin D intakes in countries where fortification is recommended or mandatory; in the US, vitamin D-fortified milk makes the highest contribution to vitamin D intakes (58% in men, 39% in women), while in other countries without fortification policies, the supply of vitamin D with milk is quite low [47].

Generally speaking, the results of surveys on representative samples of the general population in Europe suggest that dietary vitamin D intake is low overall and, on average, far from even the 5 µg set as a daily reference value. The EPIC (European Prospective Investigation into Cancer and Nutrition) study showed that vitamin D content is different in diets of people living in the 10 countries involved in the project, with intakes below mean values recorded among men and women recruited in Italy (over half of the European mean intake) and France, and intakes above the EPIC mean in Scandinavian countries (Norway and Sweden) [57]. A similar observation, namely the geographical gradient of dietary vitamin D from Northern to Southern Europe, has been described by other authors and also for children [47,58] (Figure 2). This variability can be explained by the higher consumption in Nordic countries of food products that are frequently fortified with vitamin D, where they play an important role in enabling the general population to meet the vitamin requirement.

In particular, milk is among the foods that are most commonly supplemented with vitamin D, to enhance the adequate intake of the vitamin in the general population, being of common and daily consumption for a large part of any population around the world [59].

In fact, cow's milk (and dairy products in general) has been part of the human diet for millennia, being characterized by a favorable nutrient profile; it is composed of about 87% water and also contains, on average, 3.5% protein, about 5% lactose, an amount of fat variable from 0.5% (in skimmed milk) to 3–4% (in whole milk) and minerals (1.2% mainly calcium and phosphorus) [60]. However, the content of liposoluble vitamins is low and that of vitamin D can be negligible in raw milk, as well as in yogurt [12].

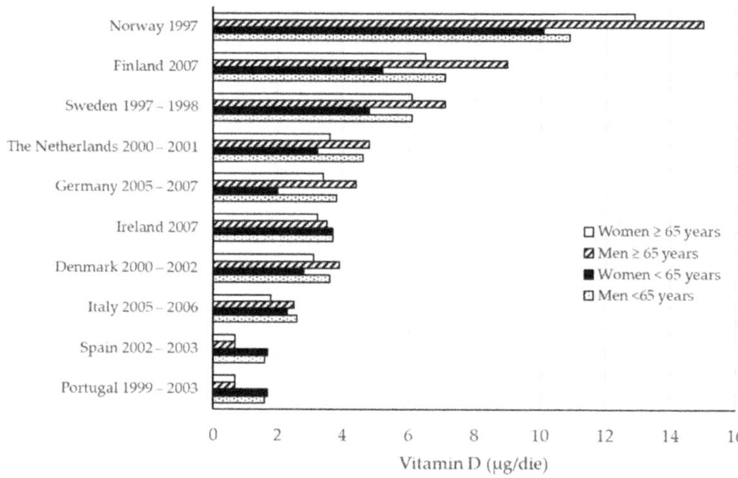

Figure 2. Vitamin D intake (μg/day) in adults living in some European countries (modified from [47]).

6. Milk as an Optimal Carrier of Vitamin D

Milk is a tremendously complex system from a physico-chemical point of view. The bio-synthetic processes involved in milk secretion require the mammary cells to perform an uninterrupted work of filtration of precursors and water from blood stream [61]. It has been estimated that more than 500 L of blood have to flow through the udder to produce one liter of milk. Within the mammary epithelial cells, precursors are constantly assembled into milk components, some of which have to be further organized in specific supramolecular structures with a high degree of complexity, i.e., casein micelles and fat globules. It has been recently suggested that these structures are envisioned in milk to act as natural nanocarriers of biologically active molecules that can thus be delivered to the neonate [62]. The comprehension of this intriguing perspective needs a discussion in light of the most recent acquisitions regarding selected properties of protein and fat structures that make them suitable to play that role.

6.1. Casein Micelles and Whey Proteins

Milk proteins comprise two main groups: Casein and whey proteins. Casein represents 80% of milk proteins and is present in a colloidal form, namely micelles, that gives the milk the white appearance. Around 10^{14}–10^{16} micelles are present in one milliliter of milk. Despite the number of studies carried out even in recent years, the structure of casein micelles is still partly unknown. Casein is actually a family of four distinct proteins, i.e., β-, αs1-, αs2- and κ-casein, with a scarcely organized structure and a strong tendency to self-associate into micelles through interactions of hydrophobic segments in the protein chains, hydrogen bonds, ionic bonds and van der Waals interactions. This architecture is stabilized by nanoclusters of calcium phosphate embedded within the protein matrix. Furthermore, the less hydrophobic κ-casein molecules are mainly positioned on the periphery of the micelle and partly extend outward forming a strongly hydrated brush layer on the surface. The overall features of casein micelles have been largely consolidated by research studies conducted over decades. Microscopy techniques such as transmission electron microscopy give evidence of a roughly spherical shape (Figure 3), and a size distribution roughly ranging from 50 to 500 nm is confirmed by light scattering measurements. How-ever, several aspects have not yet been fully elucidated. Among these, the relative distribution of water and protein throughout the micelle is still under debate. Micelles are highly hydrated, with 3–3.5 g water per g protein, and 55–60% of water associated with the micelle is entrapped in the inner part [63]. Recent studies assign the casein micelle a sponge-like structure, with many pores on the surface and channels in the core portion. The presence of numerous cavities 20–30 nm wide has been observed, filled of water and interconnected by irregular channels of about 5 nm in diameter [64]. While functional and technological behavior of casein typically depends on the surface properties of the micelles and the environmental conditions (pH, temperature, ionic concentration), the inner structural features are likely relevant in allowing the casein micelle to act as a carrier of bioactive compounds. In particular, it is emerging that lipophilic bioactives, like vitamin D, can be carried also by casein micelles besides fat globules [65]. Owing to the sponge-like structure of the micelle, these compounds can be involved in dynamic exchanges, i.e., retention into the micelle and release, depending on their characteristics.

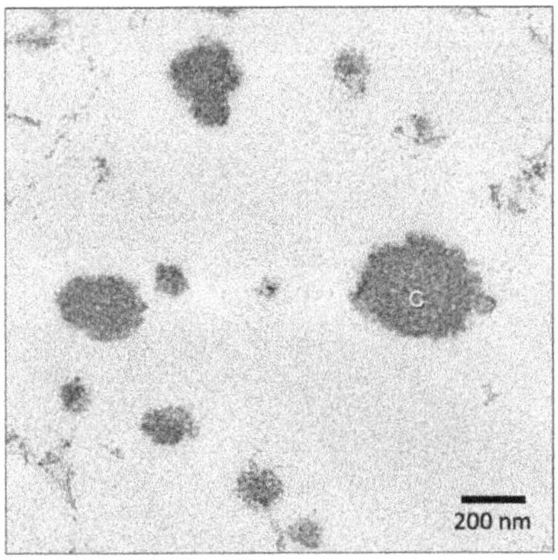

Figure 3. Transmission electron microscopy of casein micelles (C) in raw milk. Sample was prepared as described by D'Incecco et al. 2018 [66].

Globular proteins such as milk whey proteins also have the capability of carrying hydrophobic compounds by surface interaction, but the ligand can also enter the cavity of the protein. The native β-lactoglobulin is a good carrier of vitamin D, retinol and fatty acids [67] (pp. 211–259), whereas α-lactalbumin has a specific site for binding vitamin D [68]. Blood serum albumin binds vitamins, hormones and nucleotides. Furthermore, bovine milk contains the vitamin-D binding protein, a specific carrier glycoprotein with molecular characteristics close to the homologous protein in human milk [69].

6.2. Fat Globules

Fat globules are lipid droplets 0.1–10 μm in size, naturally occurring in milk as a fine emulsion. Fat globules are surrounded by a complex membrane, 8–10 nm thick, comprised of phospholipids, proteins, sterols and enzymes [70]. The membrane prevents fat globule coalescence and flocculation, thus providing stability to the emulsion phase [71]. The lipid core of the globule is primarily a mixture of triglycerides, but it also comprises a variety of lipid-soluble compounds that are entrapped together with the fat during globule assembling in the mammary cell. The lamellar distribution of triglycerides within the globule as well as the globule membrane are well visible by transmission electron microscopy (Figure 4). The presence of compounds such as vitamin D, retinol (vitamin A) and β-carotene contributes to protect milk fat from oxidation [62]. It has been demonstrated that these compounds move through the fat globule membrane by a passive diffusion process. Thus, an additional amount of vitamin D can be loaded into the native fat globules by permeation to the lipid core through the membrane [72] that, in turn, protects the vitamin D against the acidic conditions of the gastric environment. Studies on gastrointestinal digestion of milk indicate that the presence of membrane fragments in the gut lumen promotes the interaction with bile [73].

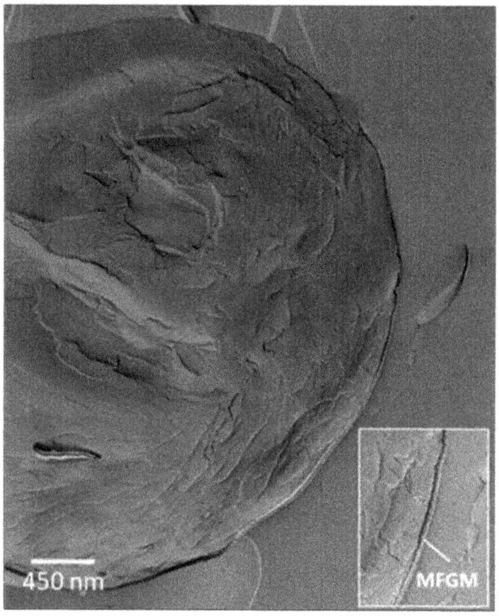

Figure 4. Freeze-fracturing transmission electron microscopy of a milk fat globule in raw milk and (box) milk fat globule membrane (MFGM).

6.3. *Industrial Processes and Storage of Milk*

The impact of industrial processing and storage of consumption milk on the stability of vitamin D determines the actual intake of this compound. Overall, it has been reported that vitamin D3 in dairy products is more stable than the other forms [8]. While vitamin D itself is prone to degradation by heat, light and oxygen, it becomes very stable when protected by food matrix. Differently, the processing and storage conditions induce deep changes to milk components, namely protein and fat.

Raw milk refrigeration is a common practice during transportation and at the processing plants. However, storage at low temperature has been shown to result in slow release of phospholipids from the fat globule membrane that becomes progressively thinner and weaker [70]. A similar structure weakening involves casein micelles due to the cold-induced release of β-casein and calcium phosphate.

Homogenization is also a common step in the industrial manufacture of drinking milk. Basically, it is a high-pressure treatment aimed at mechanically disrupting the fat globules into smaller (<1 μm) particles that are more stable against creaming. The surface of newly formed globules may increase up to 20–30 times [74] and thus casein micelles and whey protein molecules are adsorbed at the surface to rebuild an interface between the fat core and the aqueous environment outside [66] (Figure 5). Though very important for fat digestion, milk homogenization alters both the permeability of the membrane and the surface activity of fat globules compared to the native status.

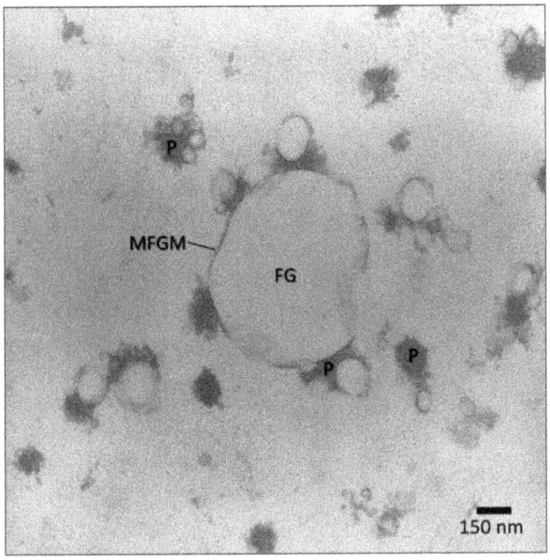

Figure 5. Transmission electron microscopy of fat globules (FG) and protein (P) in homogenized pasteurized milk. The sample was prepared as described by D'Incecco et al. 2018 [66]. MFGM = milk fat globule membrane.

Thermal treatments are used to destroy spoilage microorganisms, including any pathogens, and to preserve milk quality during an adequate shelf-life period. Vitamin D is known to be stable to thermal treatments commercially adopted by the dairy industry, i.e., pasteurization and UHT sterilization [75]. In contrast, changes induced by heating milk components, like those induced by homogenization, have an indirect effect on their capability to act as a carrier of micronutrients including vitamins. Depending on the process intensity, progressively more whey proteins are denatured and irreversibly interact with both casein micelles and proteins of fat globule membrane. It should also be mentioned that the current health trends promote the consumption of low-fat dairy products. In these products, mostly proteins act as nanocarriers for vitamin D and other hydrophobic nutraceuticals. Due to the increasing demand for protein-based nanocarriers for the delivery of these compounds, structures have been developed that mimic the behavior of native casein micelles [65].

7. Fortification Policies

All the described characteristics of milk, making it an optimal carrier for vitamin D, underlie the decision of some governments to encourage, or in some cases to compel, companies producing specific categories of foods to add vitamin D to a significant extent, on the basis of a solid nutritional rationale [14]. This is the case of foods, commonly containing very low amounts of the vitamin, that are consumed regularly and on a daily basis, which can provide, if fortified, a steady and nutritionally relevant supply of the vitamin [76]. In particular, the fortification of milk (as well as spreadable fats) is strongly recommended in Finland (1 µg/100 g in milk, yogurt and sour milk), while in Sweden, fortification is mandatory for milk with a fat content below 3% by weight (as well as for vegetable drinks intended for use as alternatives to milk: 0.38–0.50 µg/100 g). A similar approach has been adopted by Canada (milk: 0.825–1.125 µg/100 g) [14].

The safety of consumption of vitamin D-fortified foods has been confirmed by several studies [77]. Data from surveys carried out in 14 European countries indicate that vitamin D intakes from all sources, including fortified foods, within commonly consumed diets, are far below the upper tolerable levels for both adults and children [54].

The effectiveness of policies aimed at improving intakes of vitamin D by the general population through food fortification, whose adoption is relatively recent, has begun to be reported in recent years. In Canada, the fortification of milk, yogurt and cheese at 6.75 μg (270 IU)/serving led to more than doubling the vitamin D intakes across all sex/age groups and a drop in the prevalence of dietary inadequacy, from >80% to <50%, without a risk of excess [78]. Similarly, the improvement of vitamin D status in the Finnish adult population between 2000 and 2011 has been mainly explained by food fortification, especially of fluid milk products (whose contribution to dietary vitamin D intake changed from 4 to 34% in the same period), other than augmented vitamin D supplement use [79]. At the same time, the risk of insufficient vitamin D levels was reduced in population groups consuming fortified milk on a regular basis in comparison with those excluding this food from the diet, of both adults and children: Milk fortified with vitamin D was the main dietary source of vitamin D, providing 48.7% of daily intake in a population of Finnish children [80].

The study of other health effects of policies aimed at improving vitamin D intakes through food fortification is still ongoing. For example, in Denmark, prenatal exposure to small amounts of extra vitamin D from food fortification has been associated with a lower risk of developing inflammatory bowel disease before 30 years of age [81].

8. Conclusions

The critical evaluation of the relevant literature on clinical, nutritional and technological aspects allows to highlight the following points: (1) Vitamin D is essential in bone remodeling throughout the lifespan, with some specificities at each age, and for the homeostasis of many other essential metabolic processes; (2) vitamin D deficiency, which is widespread worldwide, potentially causes serious adverse effects such as rickets, osteomalacia or osteoporosis with a high risk of bone fractures and impacts on the functionality and maintenance of many physiological systems; (3) the risk of vitamin D deficiency has been described in different population groups, even in people living in sunny regions; (4) the contribution of the diet to the daily amount of vitamin D is overall modest, in the absence of fortified foods; (5) the fortification of foods consumed on a daily basis has shown to be effective in improving the intake of the vitamin in the general population; (6) in particular, the fortification of cow's milk with vitamin D is strongly recommended by some governments; (7) the assessment of the unique compositional characteristics of milk allows to conclude that it represents a good carrier for vitamin D, able to provide a small but significant amount of vitamin D, also contributing to reaching the adequate daily intake.

Author Contributions: Conceptualization, L.P., F.M. and A.C.; writing—original draft preparation, L.P., F.M., G.M., P.D., G.T.D., C.A. and A.C.; writing—review and editing, L.P. and F.M. All authors have read and agreed to the published version of the manuscript.

Funding: This research received no external funding.

Acknowledgments: Microscopy observations were carried out at The Advanced Microscopy Facility Platform—UNItech NOLIMITS—University of Milan.

Conflicts of Interest: The authors declare no conflict of interest.

References

1. Cashman, K.D.; Dowling, K.G.; Škrabáková, Z.; Gonzalez-Gross, M.; Valtueña, J.; De Henauw, S.; Moreno, L.; Damsgaard, C.T.; Michaelsen, K.F.; Mølgaard, C.; et al. Vitamin D deficiency in Europe: Pandemic? *Am. J. Clin. Nutr.* **2016**, *103*, 1033–1044. [CrossRef]

2. Liu, X.; Baylin, A.; Levy, P.D. Vitamin D deficiency and insufficiency among US adults: Prevalence, predictors and clinical implications. *Br. J. Nutr.* **2018**, *119*, 928–936. [CrossRef]
3. Lips, P.; Cashman, K.D.; Lamberg-Allardt, C.; Bischoff-Ferrari, H.A.; Obermayer-Pietsch, B.; Bianchi, M.L.; Stepan, J.; El-Hajj Fuleihan, G.; Bouillon, R. Current vitamin D status in European and Middle East countries and strategies to prevent vitamin D deficiency: A position statement of the European Calcified Tissue Society. *Eur. J. Endocrinol.* **2019**, *180*, 23–54. [CrossRef] [PubMed]
4. EFSA, Panel on Dietetic Products, Nutrition and Allergies (NDA). Scientific opinion on the substantiation of health claims related to vitamin D and normal function of the immune system and inflammatory response (ID 154, 159), maintenance of normal muscle function (ID 155) and maintenance of normal cardiovascular function (ID 159) pursuant to Article 13(1) of Regulation (EC) No 1924/2006. *EFSA J.* **2010**, *8*, 1468. [CrossRef]
5. Institute of Medicine. *Dietary Reference Intakes for Calcium and Vitamin D*; National Academies Press: Washington, DC, USA, 2010.
6. EFSA, Panel on Dietetic Products, Nutrition and Allergies (NDA). Scientific opinion on the substantiation of a health claim related to vitamin D and contribution to the normal function of the immune system pursuant to Article 14 of Regulation (EC) No 1924/2006. *EFSA J.* **2015**, *13*, 4096. [CrossRef]
7. Scientific Advisory Committee on Nutrition (SACN). Vitamin D and Health Report Public Health England. 2016. Available online: https://www.gov.uk/government/publications/sacn-vitamin-d-and-health-report (accessed on 21 June 2021).
8. Zahedirad, M.; Asadzadeh, S.; Nikooyeh, B.; Neyestani Tirang, R.; Khorshidian, N.; Yousefi, M.; Mortazavian Amir, M. Fortification aspects of vitamin D in dairy products: A review study. *Int. Dairy J.* **2019**, *94*, 53–64. [CrossRef]
9. International Dairy Federation (IDF) Bulletin N° 506/2020: The World Dairy Situation. 2020. Available online: https://fil-idf: newsinsights/idf-world-dairy-situation-report-2020/ (accessed on 21 June 2021).
10. FAO, Food and Agriculture Organization of the United Nations. Gateway to Dairy Production and Products. 2018. Available online: http://www.fao.org/dairy-production-products/production/en/ (accessed on 21 June 2021).
11. International Dairy Federation (IDF). Dairy's Role in Healthy and Sustainable Diets (Version 6 March 2020). Available online: https://fil-idf.org/wp-content/uploads/2020/03/IDF-position-to-FAO-WHO-guiding-principles-on-sustainable-healthy-diets.pdf (accessed on 21 June 2021).
12. Schmid, A.; Walther, B. Natural Vitamin D Content in Animal Products. *Adv. Nutr.* **2013**, *4*, 453–462. [CrossRef]
13. Kasalová, E.; Aufartová, J.; Kujovská Krčmová, L.; Solichová, D.; Solich, P. Recent trends in the analysis of vitamin D and its metabolites in milk—A review. *Food Chem.* **2015**, *171*, 177–190. [CrossRef]
14. Itkonen, S.; Erkkola, M.; Lamberg-Allardt, C. Vitamin D Fortification of Fluid Milk Products and Their Contribution to Vitamin D Intake and Vitamin D Status in Observational Studies—A Review. *Nutrients* **2018**, *10*, 1054. [CrossRef]
15. Annweiler, C.; Kabeshova, A.; Callens, A.; Paty, M.L.; Duval, G.T.; Holick, M.F. Self-administered Vitamin D Status Predictor: Older adults are able to use a self-questionnaire for evaluating their vitamin D status. *PLoS ONE* **2017**, *11*, 0186578. [CrossRef]
16. Muscogiuri, G.; Altieri, B.; Annweiler, C.; Balercia, G.; Pal, H.B.; Boucher, B.J.; Cannell, J.J.; Foresta, C.; Grübler, M.R.; Kotsa, K.; et al. Vitamin D and chronic diseases: The current state of the art. *Arch. Toxicol.* **2017**, *91*, 97–107. [CrossRef]
17. Christakos, S.; Ajibade, D.V.; Dhawan, P.; Fechner, A.J.; Mady, L.J. Vitamin D: Metabolism. *Endocrinol. Metab. Clin. N. Am.* **2010**, *39*, 243–253. [CrossRef] [PubMed]
18. Muscogiuri, G. Introduction to Vitamin D: Current evidence and future directions. *Eur. J. Clin. Nutr.* **2020**, *74*, 1491–1492. [CrossRef] [PubMed]
19. Grant, W.B.; Anouti, F.; Moukayed, M. Targeted 25-hydroxyvitamin D concentration measurements and vitamin D3 supplementation can have important patient and public health benefits. *Eur. J. Clin. Nutr.* **2020**, *74*, 366–376. [CrossRef] [PubMed]
20. Lips, P. Vitamin D deficiency and secondary hyperparathyroidism in the elderly: Consequences for bone loss and fractures and therapeutic implications. *Endocr. Rev.* **2001**, *22*, 477–501. [CrossRef]
21. Bover, J.; Egido, J.; Fernández-Giráldez, E.; Praga, M.; Solozábal-Campos, C.; Torregrosa, J.V.; Martínez-Castelao, A. Vitamin D, vitamin D receptor and the importance of its activation in patients with chronic kidney disease. *Rev. Nefrol.* **2015**, *35*, 28–41. [CrossRef]
22. Lips, P.; van Schoor, N.M. The effect of vitamin D on bone and osteoporosis. *Best Pract. Res. Clin. Endocrinol. Metab.* **2011**, *25*, 585–591. [CrossRef]
23. Parfitt, A.M.; Rao, D.S.; Stanciu, J.; Villanueva, A.R.; Kleerekoper, M.; Frame, B. Irreversible bone loss in osteomalacia. Comparison of radial photon absorptiometry with iliac bone histomorphometry during treatment. *J. Clin. Investig.* **1985**, *76*, 2403–2412. [CrossRef]
24. Ferrari, S.; Rizzoli, R.; Chevalley, T.; Slosman, D.; Eisman, J.A.; Bonjour, J.P. Vitamin-D- receptor-gene polymorphisms and change in lumbar-spine bone mineral density. *Lancet* **1995**, *345*, 423–424. [CrossRef]
25. Allgrove, J.; Shaw, N.J.; Allgrove, J.; Shaw, N.J. A Practical Approach to Vitamin D Deficiency and Rickets. In *Calcium and Bone Disorders in Children and Adolescents*; Allgrove, J., Shaw, N.J., Eds.; Endocrine Development Karger: Basel, Switzerland, 2015; Volume 28, pp. 119–133. [CrossRef]
26. National Institute for Health and Care Excellence (NICE). *Vitamin D: Supplement Use in Specific Population Groups*; Public Health Guideline (PH56); National Institute for Health and Care Excellence (NICE): London, UK, 2017.

27. National Health Service (NHS). Treatment of Vitamin D Deficiency in Adults. Reviewed June 2020. Available online: https://mm.wirral.nhs.uk/document_uploads/guidelines/Vitamin%20D%20Guidelines%20for%20Adults%20v2.pdf (accessed on 21 June 2021).
28. Schilling, S. Epidemic vitamin D deficiency among patients in an elderly care rehabilitation facility. *Dtsch. Arztebl. Int.* **2012**, *109*, 33–38. [CrossRef]
29. Holick, M.F.; Matsuoka, L.Y.; Wortsman, J. Age, vitamin D, and solar ultraviolet. *Lancet* **1989**, *334*, 1104–1105. [CrossRef]
30. Duval, G.T.; Rolland, G.Y.; Schott, A.M.; Blain, H.; Dargent-Molina, P.; Walrand, S.; Duque, G.; Annweiler, C. Association of hypovitaminosis D with triceps brachii muscle fatigability among older women: Findings from the EPIDOS cohort. *Maturitas* **2018**, *111*, 47–52. [CrossRef]
31. Bischoff-Ferrari, H.A.; Dawson-Hughes, B.; Staehelin, H.B.; Orav, J.E.; Stuck, A.E.; Theiler, R.; Wong, J.B.; Egli, A.; Kiel, D.P.; Henschkowski, J. Fall prevention with supplemental and active forms of vitamin D: A meta-analysis of randomised controlled trials. *BMJ* **2009**, *339*, b3692. [CrossRef]
32. Holick, M.F. Resurrection of vitamin D deficiency and rickets. *J. Clin. Investig.* **2006**, *116*, 2062–2072. [CrossRef]
33. Rolland, Y.; Mathieu, C.; Piau, C.; Cayla, F.; Bouget, C.; Vellas, B.; De Souto Barreto, P. Improving the Quality of Care of Long-Stay Nursing Home Residents in France. *J. Am. Geriat. Soc.* **2016**, *64*, 193–199. [CrossRef]
34. Bischoff-Ferrari, H.A.; Willett, W.C.; Wong, J.B.; Giovannucci, E.; Dietrich, T.; Dawson-Hughes, B. Fracture prevention with vitamin D supplementation: A meta-analysis of randomized controlled trials. *JAMA* **2005**, *293*, 2257–2264. [CrossRef]
35. Bischoff-Ferrari, H.A.; Willett, W.C.; Wong, J.B.; Stuck, A.E.; Staehelin, H.B.; Orav, E.J.; Thoma, A.; Kiel, D.P.; Henschkowski, J. Prevention of nonvertebral fractures with oral vitamin D and dose dependency: A meta-analysis of randomized controlled trials. *Arch. Intern. Med.* **2009**, *169*, 551–561. [CrossRef]
36. Annweiler, C.; Beauchet, O. Questioning vitamin D status of elderly fallers and nonfallers: A meta-analysis to address a 'forgotten step'. *J. Intern. Med.* **2015**, *277*, 16–44. [CrossRef] [PubMed]
37. Yao, P.; Bennett, D.; Mafham, M.; Lin, X.; Chen, Z.; Armitage, J.; Clarke, R. Vitamin D and calcium for the prevention of fracture: A systematic review and meta-analysis. *JAMA Netw. Open* **2019**, *2*, e1917789. [CrossRef] [PubMed]
38. Reid, I.R.; Bolland, M.J.; Grey, A. Effects of vitamin D supplements on bone mineral density: A systematic review and meta-analysis. *Lancet* **2014**, *383*, 146–155. [CrossRef]
39. Muscogiuri, G.; Mitri, J.; Mathieu, C.; Badenhoop, K.; Tamer, G.; Orio, F.; Mezza, T.; Vieth, R.; Colao, A.; Pittas, A. Mechanisms in endocrinology: Vitamin D as a potential contributor in endocrine health and disease. *Eur. J. Endocrinol.* **2014**, *171*, R101–R110. [CrossRef]
40. Altieri, B.; Grant, W.B.; Della Casa, S.; Orio, F.; Pontecorvi, A.; Colao, A.; Sarno, G.; Muscogiuri, G. Vitamin D and pancreas: The role of sunshine vitamin in the pathogenesis of diabetes mellitus and pancreatic cancer. *Crit. Rev. Food Sci. Nutr.* **2017**, *57*, 3472–3488. [CrossRef] [PubMed]
41. Muscogiuri, G.; Barrea, L.; Scannapieco, M.; Di Somma, C.; Scacchi, M.; Aimaretti, G.; Savastano, S.; Colao, A.; Marzullo, P. The lullaby of the sun: The role of vitamin D in sleep disturbance. *Sleep Med.* **2019**, *54*, 262–265. [CrossRef] [PubMed]
42. Charoenngam, N.; Holick, M.F. Immunologic effects of vitamin D on human health and disease. *Nutrients* **2020**, *12*, 2097. [CrossRef] [PubMed]
43. Schwalfenberg, G.K. A review of the critical role of vitamin D in the functioning of the immune system and the clinical implications of vitamin D deficiency. *Mol. Nutr. Food Res.* **2011**, *55*, 96–108. [CrossRef]
44. Roth, D.E.; Abrams, S.A.; Aloia, J.; Bergeron, G.; Bourassa, M.W.; Brown, K.H.; Calvo, M.S.; Cashman, K.D.; Combs, G.; De-Regil, L.M.; et al. Global prevalence and disease burden of vitamin D deficiency: A roadmap for action in low- and middle-income countries. *Ann. N. Y. Acad. Sci.* **2018**, *1430*, 44–79. [CrossRef] [PubMed]
45. ODIN (Food-Based Solutions for Optimal Vitamin D Nutrition and Health through the Life Cycle). Available online: https://cordis.europa.eu/project/id/613977/reporting/it (accessed on 5 June 2021).
46. Wahl, D.A.; Cooper, C.; Ebeling, P.R.; Eggersdorfer, M.; Hilger, J.; Hoffmann, K.; Josse, R.; Kanis, J.A.; Mithal, A.; Pierroz, D.D. et al. A global representation of vitamin D status in healthy populations. *Arch. Osteoporos.* **2012**, *7*, 155–172. [CrossRef]
47. Spiro, A.; Buttriss, J.L. Vitamin D: An overview of vitamin D status and intake in Europe. *Nutr. Bull.* **2014**, *39*, 322–350. [CrossRef]
48. Van Schoor, N.; Lips, P. Global Overview of Vitamin D Status. *Endocrinol. Metab. Clin. N. Am.* **2017**, *46*, 845–870. [CrossRef]
49. Karras, S.N.; Anagnostis, P.; Annweiler, C.; Naughton, D.P.; Petroczi, A.; Bili, E.; Harizopoulou, V.; Tarlatzis, B.C.; Persinaki, A.; Papadopoulou, F.; et al. Maternal vitamin D status during pregnancy: The Mediterranean reality. *Eur. J. Clin. Nutr.* **2014**, *68*, 864–869. [CrossRef]
50. Braegger, C.; Campoy, C.; Colomb, V.; Decsi, T.; Domellof, M.; Fewtrell, M.; Hojsak, I.; Mihatsch, W.; Molgaard, C.; Shamir, R. et al. Vitamin D in the Healthy European Paediatric Population. *J. Pediatr. Gastroenterol. Nutr.* **2013**, *56*, 692–701. [CrossRef] [PubMed]
51. Herrick, K.A.; Storandt, R.J.; Afful, J.; Pfeiffer, C.M.; Schleicher, R.L.; Gahche, J.J.; Potischman, N. Vitamin D status in the United States, 2011–2014. *Am. J. Clin. Nutr.* **2019**, *110*, 150–157. [CrossRef]
52. Calame, W.; Street, L.; Hulshof, T. Vitamin D Serum Levels in the UK Population, including a Mathematical Approach to Evaluate the Impact of Vitamin D Fortified Ready-to-Eat Breakfast Cereals: Application of the NDNS Database. *Nutrients* **2020**, *12*, 1868. [CrossRef] [PubMed]

53. Regulation (EU) No 1169/2011 of the European Parliament and of the Council of 25 October 2011 on the provision of food information to consumers, amending Regulations (EC) No 1924/2006 and (EC) No 1925/2006 of the European Parliament and of the Council, and repealing Commission Directive 87/250/EEC, Council Directive 90/496/EEC, Commission Directive 1999/10/EC, Directive 2000/13/EC of the European Parliament and of the Council, Commission Directives 2002/67/EC and 2008/5/EC and Commission Regulation (EC) No 608/2004. *Off. J. Eur. Un. L.* **2011**, *304*, 18–63.
54. EFSA, Panel on Dietetic Products, Nutrition and Allergies (NDA). Scientific Opinion on the Tolerable Upper Intake Level of vitamin D. *EFSA J.* **2012**, *10*, 2813. [CrossRef]
55. Bresson, J.L.; Burlingame, B.; Dean, T.; Fairweather-Tait, S.; Heinonen, M.; Hirsch-Ernst, K.I.; Mangelsdorf, I.; McArdle, H.; Naska, A.; Neuhäuser-Berthold, M.; et al. Dietary reference values for vitamin D. *EFSA J.* **2016**, *14*, e04547. [CrossRef]
56. Giustina, A.; Adler, R.A.; Binkley, N.; Bollerslev, J.; Bouillon, R.; Dawson-Hughes, B.; Ebeling, P.R.; Feldman, D.; Formenti, A.M.; Lazaretti-Castro, M.; et al. Consensus statement from 2nd International Conference on Controversies in Vitamin D. *Rev. Endocr. Metab. Disord.* **2020**, *21*, 89–116. [CrossRef]
57. Freisling, H.; Fahey, M.T.; Moskal, A.; Ocké, M.C.; Ferrari, P.; Jenab, M.; Norat, T.; Naska, A.; Welch, A.A.; Navarro, C.; et al. Region-specific nutrient intake patterns exhibit a geographical gradient within and between European countries. *J. Nutr.* **2010**, *140*, 1280–1286. [CrossRef]
58. Elmadfa, I.; Meyer, A.; Nowak, V.; Hasenegger, V.; Putz, P.; Verstraeten, R.; Remaut-DeWinter, A.M.; Kolsteren, P.; Dostálová, J.; Dlouhý, P.; et al. European Nutrition and Health Report 2009. *Forum Nutr.* **2009**, *62*, 1–405. [CrossRef]
59. Lavelli, V.; D'Incecco, P.; Pellegrino, L. Vitamin D incorporation in foods: Formulation strategies, stability and bioaccessibility as affected by the food matrix. *Foods* **2021**, *10*, 1989. [CrossRef]
60. Marangoni, F.; Pellegrino, L.; Verduci, E.; Ghiselli, A.; Bernabei, R.; Calvani, R.; Irene, C.; Michelangelo, G.; Francesco, P.; Luca, P.; et al. Cow's milk consumption and health: A health professional's guide. *J. Am. Coll. Nutr.* **2019**, *38*, 197–208. [CrossRef]
61. Cai, J.; Wang, D.; Liu, J. Regulation of fluid flow through the mammary gland of dairy cows and its effect on milk production: A systematic review. *J. Sci. Food Agric.* **2018**, *98*, 1261–1270. [CrossRef] [PubMed]
62. Acevedo-Fani, A.; Dave, A.; Singh, H. Nature-assembled structures for delivery of bioactive compounds and their potential in functional foods. *Front. Chem.* **2020**, *8*, 564021. [CrossRef]
63. Huppertz, T.; Gazi, I.; Luyten, H.; Nieuwenhuijse, H.; Alting, A.; Schokker, E. Hydration of casein micelles and caseinates: Implications for casein micelle structure. *Int. Dairy J.* **2017**, *74*, 1–11. [CrossRef]
64. Trejo, R.; Dokland, T.; Jurat-Fuentes, J.; Harte, F. Cryo-transmission electron tomography of native casein micelles from bovine milk. *J. Dairy Sci.* **2011**, *94*, 5770–5775. [CrossRef]
65. Cohen, Y.; Margier, M.; Lesmes, U.; Reboul, E.; Livney, Y.D.D. Mechanisms of Absorption of Vitamin D3 Delivered in Protein Nanoparticles in the Absence and Presence of Fat. *Food Funct.* **2021**, *12*, 4935–4946. [CrossRef] [PubMed]
66. D'Incecco, P.; Rosi, V.; Cabassi, G.; Hogenboom, J.A.; Pellegrino, L. Microfiltration and ultra-high-pressure homogenization for extending the shelf-storage stability of UHT milk. *Food Res. Int.* **2018**, *107*, 477–485. [CrossRef]
67. Sawyer, L. *β-Lactoglobulin, Advanced Dairy Chemistry*; Springer: Boston, MA, USA, 2013; pp. 211–259.
68. Kimpel, F.; Schmitt, J.J. Review: Milk Proteins as Nanocarrier Systems for Hydrophobic Nutraceuticals. *J. Food Sci.* **2015**, *80*, 2361–2366. [CrossRef]
69. Wynn, P.C.; Sheehy, P.A. *Minor Proteins, Including Growth Factors. Advanced Dairy Chemistry*; Springer: Boston, MA, USA, 2013; pp. 317–335.
70. Singh, H.; Gallier, S. Nature's complex emulsion: The fat globules of milk. *Food Hydrocoll.* **2017**, *68*, 81–89. [CrossRef]
71. D'Incecco, P.; Ong, L.; Pellegrino, L.; Faoro, F.; Barbiroli, A.; Gras, S. Effect of temperature on the microstructure of fat globules and the immunoglobulin-mediated interactions between fat and bacteria in natural raw milk creaming. *J. Dairy Sci.* **2018**, *101*, 2984–2997. [CrossRef] [PubMed]
72. Alshehab, M.; Reis, M.G.; Day, L.; Nitin, N. Milk fat globules, a novel carrier for delivery of exogenous cholecalciferol. *Food Res. Int.* **2019**, *126*, 108579. [CrossRef]
73. Tunick, M.H.; Ren, D.X.; Van Hekken, D.L.; Bonnaillie, L.; Paul, M.; Kwoczak, R.; Tomasula, P.M. Effect of heat and homogenization on in vitro digestion of milk. *J. Dairy Sci.* **2016**, *99*, 6. [CrossRef]
74. Obeid, S.; Guyomarc'h, F.G.; Tanguy Leconte, N.; Rousseau, F.; Dolivet, A.; Leduc, A.; Wu, X.; Cauty, C.; Jan, G.; Gaucheron, F.; et al. The adhesion of homogenized fat globules to proteins is increased by milk heat treatment and acidic pH: Quantitative insights provided by AFM force spectroscopy. *Food Res. Int.* **2020**, *129*, 108847. [CrossRef] [PubMed]
75. Hanson, A.L.; Metzger, L.E. Evaluation of increased vitamin D fortification in high-temperature, short-time–processed 2% milk, UHT-processed 2% fat chocolate milk, and low-fat strawberry yogurt. *J. Dairy Sci.* **2010**, *93*, 801–807. [CrossRef]
76. Hayes, A.; Cashman, K.D. Food-based solutions for vitamin D deficiency: Putting policy into practice and the key role for research. *Proc. Nutr. Soc.* **2017**, *76*, 54–63. [CrossRef]
77. Amrein, K.; Scherkl, M.; Hoffmann, M.; Neuwersch-Sommeregger, S.; Köstenberger, M.; Tmaya-Berisha, A.; Martucci, G.; Pilz, S.; Malle, O. Vitamin D deficiency 2.0: An update on the current status worldwide. *Eur. J. Clin. Nutr.* **2020**, *74*, 1498–1513. [CrossRef] [PubMed]
78. Shakur, Y.A.; Lou, W.; L'Abbe, M.R. Examining the effects of increased vitamin D fortification on dietary inadequacy in Canada. *Can. J. Public Health* **2014**, *105*, e127–e132. [CrossRef] [PubMed]

79. Jääskeläinen, T.; Itkonen, S.T.; Lundqvist, A.; Erkkola, M.; Koskela, T.; Lakkala, K.; Dowling, K.G.; Hull, G.L.; Kröger, H.; Karppinen, J.; et al. The positive impact of general vitamin D food fortification policy on vitamin D status in a representative adult Finnish population: Evidence from an 11-y follow-up based on standardized 25-hydroxyvitamin D data. *Am. J. Clin. Nutr.* **2017**, *105*, 1512–1520. [CrossRef]
80. Soininen, S.; Eloranta, A.M.; Lindi, V.; Venäläinen, T.; Zaproudina, N.; Mahonen, A.; Lakka, T.A. Determinants of serum 25-hydroxyvitamin D concentration in Finnish children: The Physical Activity and Nutrition in Children (PANIC) study. *Br. J. Nutr.* **2016**, *115*, 1080–1091. [CrossRef] [PubMed]
81. Duus, K.S.; Moos, C.; Frederiksen, P.; Andersen, V.; Heitmann, B.L. Prenatal and Early Life Exposure to the Danish Mandatory Vitamin D Fortification Policy Might Prevent Inflammatory Bowel Disease Later in Life: A Societal Experiment. *Nutrients* **2021**, *13*, 1367. [CrossRef]

molecules

MDPI

Article

Cow and Ewe Cheeses Made with Saffron: Characterization of Bioactive Compounds and Their Antiproliferative Effect in Cervical Adenocarcinoma (HeLa) and Breast Cancer (MDA-MB-231) Cells

Mena Ritota *, Raffaella Comitato and Pamela Manzi

Consiglio per la Ricerca in Agricoltura e L'analisi Dell'economia Agraria, Centro di Ricerca Alimenti e Nutrizione, Via Ardeatina 546, 00178 Rome, Italy; raffaella.comitato@crea.gov.it (R.C.); pamela.manzi@crea.gov.it (P.M.)
* Correspondence: mena.ritota@crea.gov.it

Abstract: Saffron is a widespread consumed spice containing many phytochemicals. It is often used in dairy technologies to enhance color and flavor of cheeses, but it is also known for its several therapeutic effects, as well as its antiproliferative and anticancer properties. In this study High Performance Liquid Chromatography was used to characterize saffron bioactive compounds in cow and ewe cheeses made with saffron, and the antiproliferative effect of the crocin-rich extracts from cheeses was investigated on different cellular lines (CaCo2, MDA-MB-231 and HeLa) by MTT assay. Crocins were observed in all cheese samples, with the total content ranging between 0.54 and 30.57 mg *trans*-4-GG/100 g cheese, according to the different cheese making process. Picrocrocin was detected in no cheese (probably due to its degradation during cheese making), while safranal was detected only in one ewe cheese (mainly due to its high volatility). HeLa and MDA-MB-231 cells were sensitive to treatment with crocin-rich extracts from cheeses, while no effect was observed on CaCo2 cells. The chemical environment of the food matrix seems to have a great influence on the crocin antiproliferative effect: the crocin-rich extracts from cheese with both high residual N/protein and fat contents showed increased antiproliferative effect compared to pure crocin (*trans*-4-GG), but cheeses from different milk species (type of fats and proteins) could also play an important role in modulating crocin's antiproliferative effects.

Keywords: cow and ewe cheese; saffron; bioactive compounds; crocin; safranal; picrocrocin; antiproliferative effect

1. Introduction

Italy plays an important role within the dairy industry worldwide, being one of the largest producers of typical PDO (Protected Designation of Origin) cheeses. Despite the wide range of dairy products, those that hold real importance in commerce are very few in number and often limited to some main cheeses such as Parmigiano Reggiano PDO and Grana Padano PDO. Ewe and goat milk cheeses, instead, are mostly produced on a small local scale, and they are not as competitive as cow milk cheeses. For this reason, cheese makers need to produce innovative and high-quality products, and to also meet the increasing demands of consumers.

Innovation in cheese has been in part driven by the addition of several ingredients, such as colorants, spices, leaves, flavors, and aroma agents. A common practice is the giving of a natural yellow to orange color to cheeses through the addition of spices: annatto, for example, is used in Cheddar and Mimolette cheese making. More recently, the inclusion of saffron in the cheese-making process is currently an increasingly practice as well.

Saffron, a perennial herb belonging to the *Iridaceae* family, has a long history as a culinary spice, but since ancient times it has been also used as medicinal plant [1]. In the food sector it is mainly known for its color, odor, and taste. Picrocrocin (4-(β-D-glucopyranosyloxy)-2,6,6-trimethyl-1-cyclohexene-1-carboxaldehyde), a monoterpene glycoside produced from zeaxanthin degradation, is responsible for the bitter taste of saffron, while its thermal or enzymatic degradation [2] gives rise to safranal (2,6,6-trimethylcyclohexa-1,3-diene-1-carbaldehyde), a monoterpenoid to which saffron aroma is mainly related. Safranal has also been revealed to show bear high antioxidant potential, as well as cytotoxicity towards certain cancer cells in vitro [3]. The yellow-red color of saffron is instead mainly due to a class of water-soluble carotenoids, crocins, which are esters of crocetin (8,8′-diapocarotenedioic acid) with different glycosidic moieties, such as glucose, gentiobiose, and neapolitanose. Further lipophilic carotenoids, such as lycopene, α-, and β-carotene, as well as zeaxanthin, phytoene, and phytofluene have also been reported in saffron in trace amounts [4]. In addition to its use as food coloring, crocin acts as an antioxidant by quenching free radicals, protecting cells and tissues against oxidation [3]. Other phytochemical compounds of saffron also include a wide range of flavonoids (mainly kaempferol derivatives), phenolic acids (hydroxycinnamic and hydroxybenzoic acids being the most representative) and phytosterols [4].

Owing to the presence of all these bioactive compounds, several therapeutic effects have been attributed to saffron, including cardiovascular disease prevention, insulin resistance, gastric disorders, depression, insomnia, premenstrual syndrome, anxiety, and neurodegenerative disorders, as well as bearing learning and memory-enhancing effects [3–5]. In the last decades studies demonstrating the anticancer and antiproliferative properties of saffron and its extracts have been published [6–8]. Among the three main saffron bioactive compounds, crocin has revealed to contain the most effective anticancer activity [9,10], and it has been demonstrated that crocin possesses antiproliferative effects on several cell lines derived from most types of human cancers [7], operating through several molecular mechanisms [7,9].

Although there are many studies regarding saffron addition to cheese infused [11–13], very limited works are reported in the literature concerning the characterization of saffron bioactive molecules in cheese, most of them regarding the determination of the volatile profile, including safranal [14,15], and only one regarding crocins and picrocrocin evaluation in cheese [16]. Furthermore, the antiproliferative effects of saffron have been mainly evaluated in pure saffron components [17–20] or its extracts [21–23], while there is a lack of studies concerning evaluation of the antiproliferative effect of saffron extract within a food matrix or in a complex mixture, additionally, no evaluation of the influence of the food chemical environment on saffron's antiproliferative action has been reported.

Considering all these aspects, the objectives of this study were: (i) to characterize saffron bioactive molecules (crocins, picrocrocin and safranal) in cow and ewe cheeses made with saffron through a UHPLC (Ultra High Performance Liquid Chromatography) analytical method; (ii) to evaluate the in vitro antiproliferative effects of cheese extracts containing saffron bioactive compounds on different cellular lines, CaCo2 (colon cancer), HeLa (cervix carcinoma) and MDA-MB-231 (breast cancer); (iii) to evaluate any influence of the cheese matrix in modulating the antiproliferative effects of saffron's bioactive compounds.

2. Results and Discussion

2.1. *Saffron Bioactive Molecules in Cheese*

Figure 1 shows UHPLC chromatograms of saffron bioactive molecules, picrocrocin, crocins and safranal, recorded at 250, 440 and 310 nm, respectively.

Picrocrocin (peak 1 in Figure 1a) was identified according to retention time of the standard and its UV spectrum [16], as reported in Figure 2. However, this molecule, to which saffron taste is related, was observed only in saffron extract and not in cheese extracts, confirming previous data reported in the literature according to which the lack of picrocrocin in the hydroalcoholic extract of cheese is probably due to its degradation during cheese making [16].

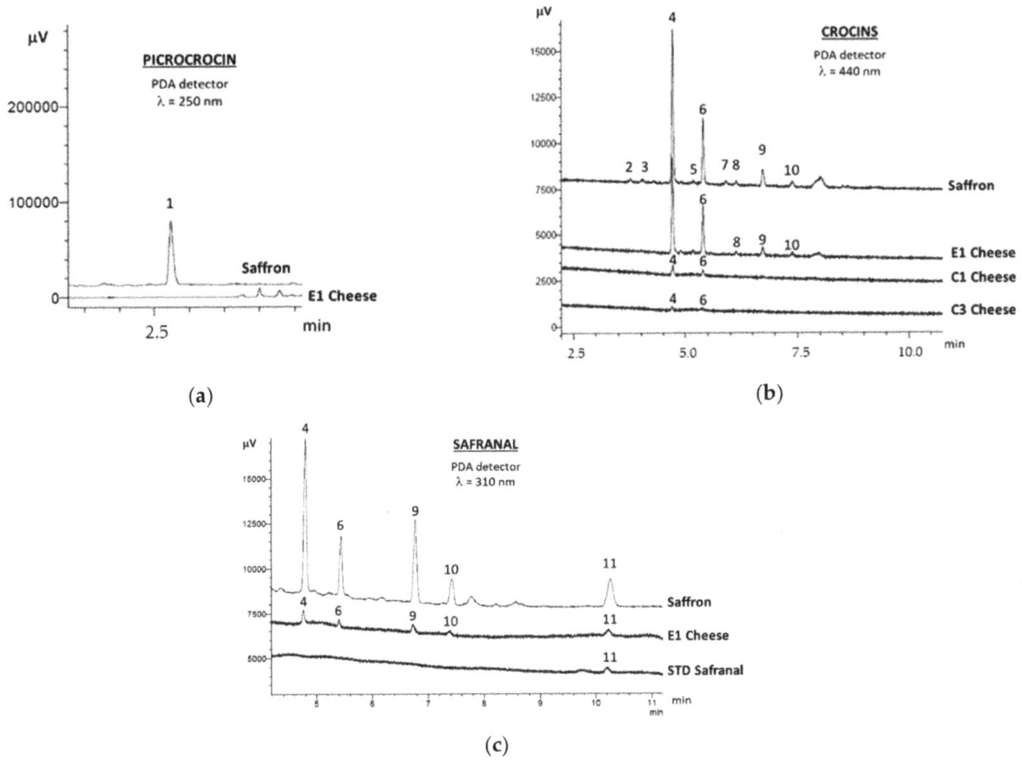

Figure 1. (**a**) UHPLC chromatograms of saffron and ewe cheese (E1) extracts detected at 250 nm; (**b**) UHPLC chromatograms of saffron, ewe (E1) and cow cheeses (C1, C3) extracts detected at 440 nm; (**c**) UHPLC chromatograms of saffron and ewe cheese (E1) extracts, together with a safranal standard (32 μg/L), detected at 310 nm: picrocrocin (peak 1); *trans*-crocins (peaks 2, 3, 5, 7, 8); *trans*-4-GG (peak 4); *trans*-3-gG (peak 6); *cis*-crocins (peaks 9, 10); safranal (peak 11).

Regarding crocins, *trans*-4-GG, was the most abundant crocin observed in all cheese samples (peak 4 in Figure 1b), ranging between 45.4 and 69.8% of the total content. It was followed by another *trans*-crocin (peak 6 in Figure 1b), tentatively assigned to *trans*-crocetin (β-D-glucosyl)-(β-D-gentibiosyl) ester (*trans*-3-Gg), according to several studies which reported *trans*-4-GG and *trans*-3-Gg as the two main crocins present in saffron hydroalcoholic extracts [24–26]. *Trans*-3-Gg ranged between 26.3 and 43.5% of the total crocins content. Two ewe cheeses (E1 in Figure 1b,c, and E3 data not shown), as well as saffron (Figure 1b,c), also showed the presence of *cis*-crocins in their chromatographic profile (peaks 9–10 in Figure 1), while two cheese samples (C2 and E2, data not shown) showed only traces of the *cis*-crocin labelled as peak 9 in Figure 1. The recognition of *trans* and *cis* crocin isomers in the chromatographic profile of cheese was possible thanks to differences in their absorption spectra (see Figure 2c,d): *trans*-crocins show two absorption bands in their UV–Vis spectrum (250–260 nm, and 400–500 nm), while the *cis* isomers have a further absorption band at around 325 nm. The presence of both *trans* and *cis*-crocins in cheeses made with saffron had been previously reported only by Ritota, et al. [16], who ascribed the content of *cis*-crocins in cheese to their natural presence in saffron added to cheese and to the *cis-trans* isomerization reaction, promoted by light, which may occur during cheese making.

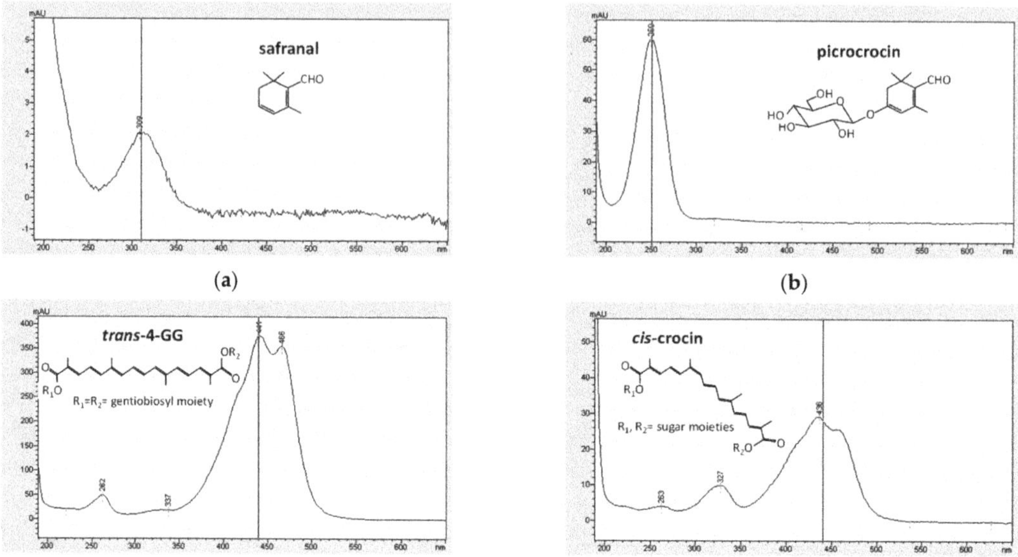

(a) (b) (c) (d)

Figure 2. UV–Vis absorption spectra and chemical structures of saffron bioactive compounds: (**a**) safranal, (**b**) picrocrocin, (**c**) *trans*-4-GG, and (**d**) *cis*-crocin. All analytes were dissolved in methanol:water (80:20, *v*/*v*).

Table 1 reports the total crocins content (sum of all *cis* and *trans*-crocins) of the cheese samples, expressed as *trans*-4-GG content. Significant differences ($p < 0.05$) are probably due to the different amounts of added saffron and/or to the different cheesemaking technologies (saffron added in stigmas or powder, to milk or curd, before or after curd cooking, etc.). However, ewe cheeses generally showed higher crocin levels compared to cow cheeses.

Table 1. Characterization of saffron bioactive compounds in cheese (mg/100 g cheese): total crocins and safranal. Different superscript letters in the same column correspond to significantly different mean values ($p < 0.05$).

Sample	Total Crocins (mg/100 g Cheese)	Safranal (μg/100 g Cheese)
E1	30.57 ± 0.49^{a}	57 ± 1
E2	2.87 ± 0.15^{c}	n.d.
E3	10.02 ± 0.12^{b}	n.d.
C1	0.83 ± 0.02^{d}	n.d.
C2	1.49 ± 0.01^{e}	n.d.
C3	0.54 ± 0.03^{d}	n.d.
C4	1.36 ± 0.09^{e}	n.d.

n.d. not detected, namely lower than safranal LOD. Different superscript letters (a–e) mean significant differences ($p < 0.05$).

The same analytical method for crocins determination was also used to characterize crocin-rich extracts from cheeses for their safranal content (Figure 1c). Safranal is a compound characterized by high volatility, hence UHPLC method would not be the suitable analytical technique for its determination since an underestimation of safranal content may occur. However, one of the aims of this work was to characterize saffron bioactive molecules in cheese through a single analytical method, and there are no data in the literature referring to safranal in cheese by UHPLC technique. As reported in Table 1, safranal was detected only in E1 ewe cheese, probably due to its high saffron content added during cheese making.

2.2. In Vitro Antiproliferative Assay of the Crocin-Rich Extracts

The antiproliferative effect of the crocin-rich extracts from cheeses was evaluated by in vitro study in three cellular models, cervix carcinoma (HeLa), colon cancer (CaCo2) and breast cancer. In order to exclude a possible interaction of estrogen receptor with the different molecules (fat residues) present in the extracts and to be able to compare the obtained results, a triple negative breast cancer line (MDA-MB-231) was used, which lacks estrogen and progesterone receptor expression, as well as human epidermal growth factor receptor 2 (HER2) amplification [27,28]. Total crocin contents of the cheese extracts were maintained constant (1 or 2 μM, according to the experimental needs) in order to evaluate any potential influence of the food matrix in modulating the antiproliferative effect of saffron bioactive compounds. Such crocin concentrations were obtained from data reported in the literature, and in particular from the work of Hire, et al. [29], who showed that crocin inhibited the assembly of microtubules, inducing a more significant antiproliferative effect in cancer cells (cervix and breast cancer cells), than non-cancerous fibroblast cells.

Figure 3 shows the results of MTT assay, where it can be observed that CaCo2 cellular model was not responsive to any experimental treatment, while significant effects were observed in HeLa and MDA-MB-231 cells.

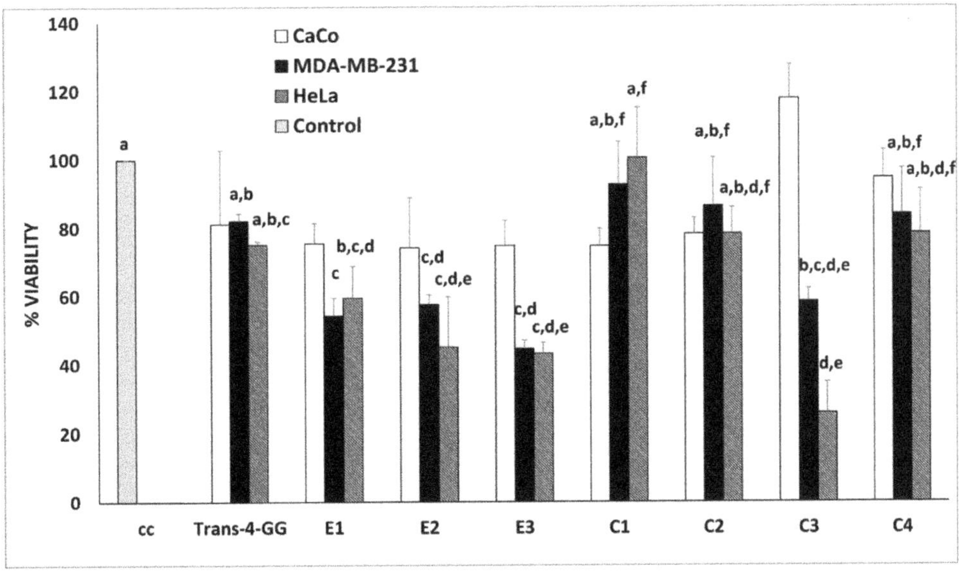

Figure 3. Antiproliferative effects (MTT assay) of a standard of *trans*-4-GG (2 μM) and of the crocin-rich extracts from cheese on CaCo2, MDA-MB-231 and HeLa cells models. Data are expressed as percentage of viability cell and presented as mean ± SE of three independent experiments, each conducted in triplicate. Different letters (a–f) mean significant differences ($p < 0.05$).

Intestinal cells show greater resistance to external insults. In fact, some authors have shown that intestinal cell models are resistant to toxins [30], and exhibit a minor sensitivity at different treatments [31,32]. We hypothesize that the greater resistance of intestinal cell (CaCo2) could be due to the higher exposure to external insults compared to other tissues. Abdullaev Jafarova, et al. [33] had already observed an inhibitory effect of saffron extracts on the colony formation of some malignant cellular lines, reporting a major sensitivity of cervix carcinoma cells (HeLa) compared to colon carcinoma cells (SW480), 82% and 63%, respectively. Breast cancer cells (MDA-MB-231,) also showed a proliferation pattern similar to HeLa cells, although the significant antiproliferative effect in some treatments was less efficient. A similar result was observed by Gezici [34], who reported a greater inhibitory effect of saffron extracts on HeLa cells compared to epithelial tumor cell lines A549 and MCF, respectively, lung and breast cancer cells.

Although the results obtained in this study highlight and corroborate the major susceptibility of HeLa and MDA-MB-231 cell model to the antiproliferative activity of the crocin-rich extracts compared to intestinal cell model (CaCo2), it is worthwhile noting that this is the first time that saffron bioactive molecules present in a food matrix (cheese extracts) are evaluated for their antiproliferative effects on cancer model cell lines.

While no statistical differences are present in CaCo2 cell lines, crocin-rich extracts from E1, E2, E3 and C3 cheeses showed a significant ($p < 0.05$) antiproliferative effect in HeLa and MDA-MB-231 cells (Figure 3). In particular, viability decrease was most evident in C3 treatment both in HeLa and MDA-MB-231 cells (~75% and 42% of cellular death, respectively).

The antiproliferative effect of the crocin-rich extracts from cheese is certainly due to saffron bioactive compounds, mainly crocins, but the potential influence of the chemical environment of food matrix on the crocin antiproliferative effect cannot be neglected since, with the same concentration of crocins in the cheese extracts, differences in the antiproliferative effects were observed.

The natural structure of whole food certainly plays an important role in the physiological activity of bioactive compounds present within the food, also known as the "matrix effect" [35–37]. The food matrix, in fact, interacts with specific compounds, providing functionalities and behaviors which are different from those exhibited by the isolate components or in free state. Some authors showed that differences among food matrices are largely responsible for the different health properties of products that share similar chemical composition [38,39]. Despite this fundamental aspect, free state nutrients are still used in most in vitro studies, with results that are not reproducible in a food or whole diet.

In this study, in order to understand if the food matrix can affect crocins antiproliferative activity, the crocin-rich extracts incubated on the cellular lines were better characterized.

2.3. Characterization of the Crocin-Rich Extracts: Residual Protein Nitrogen and Fat

The use of a methanol:water (80:20, v/v) mixture as solvent for the extraction of saffron bioactive molecules from cheese could have resulted in the extraction of small peptides and/or short-chain fatty acids, as well as polar lipids.

Biological activities of milk-derived peptides have been known for a long time [40] and among them some peptides exert cytomodulatory action which may inhibit the growth of cancer cells [41]. Regarding cheese, instead, the characterization of peptides with antiproliferative properties is still ongoing [42]. The literature demonstrated the antiproliferative effect of methanolic extracts from cheeses on human leukemia cell lines [43], the inhibitory growth activities of water-soluble peptides extract of Cheddar cheeses on different cellular cancer models [44–46], and the antiproliferative effect on CaCo2 exerted by Mozzarella di Bufala waste whey peptides [47].

Among milk fat compounds, specific fatty acids, such as butyric acid, conjugated linoleic acid (CLA), phospholipids and sphingolipids from milk globule membrane (MGM) are considered to exert potential anticancer activity [48]. Yasuda, et al. [49], for example, observed that methanolic extracts from goat cheeses with different lipid contents exerted antiproliferative activity against human leukemia cells, as well as isolates from cream and buttermilk which showed antiproliferative effects against colon cancer cells both in vitro and in vivo [50,51].

Therefore, the hydroalcoholic extracts from cheese incubated on the cellular lines were characterized for their residual N/protein and fat, together with the "normalized" crocin content, namely the exact concentration of crocins incubated on cells (Table 2), in order to evaluate the potential influence of the chemical environment (food matrix effect) on the crocin antiproliferative effect.

Table 2. Main characterization of the crocin-rich extracts incubated on the cellular lines: total crocins (μM) incubated; residual protein nitrogen (mg/mL) and fat (mg/mL); *trans*-3-Gg/*trans*-4-GG ratio and *trans*-4-GG percentage.

Sample	Total Crocins (μM)	Protein N Residue (mg/mL)	Fat Residue (mg/mL)	*trans*-3-Gg/*trans*-4-GG	*trans*-4-GG (% on Total Crocins)
E1	2.00	2.52	0.35	0.5	57.3
E2	2.00	12.88	4.16	0.4	60.3
E3	2.00	10.98	1.53	0.7	46.4
C1	1.00	76.72	8.12	0.7	58.6
C2	2.00	77.66	7.72	0.5	56.1
C3	1.00	109.01	12.96	0.4	69.4
C4	1.00	44.58	5.36	0.6	63.8

Among ewe cheeses, with equal crocin contents incubated on the cells, the highest antiproliferative effects (E2 = E3 > E1, as reported in Figure 3) were observed in the samples with both high residual N/protein and fat contents.

Cow cheeses followed the same trend observed for ewe cheeses: C3 sample, which showed the highest antiproliferative effect, also had the highest residual N/protein and fat contents.

Despite the few samples analyzed in this study, Pearson's correlation analysis showed results worth reporting (Table 3): with increasing N/protein in the crocin-rich extracts, a higher antiproliferative effect on HeLa cells was observed in ewe cheeses ($r = 0.961$), while a lower influence was observed for cow samples ($r = 0.675$, rising to 0.686 by excluding C2 sample which had a double crocins content compared to other cow cheeses). A similar result was also observed on MDA-MB-231 cell proliferation, but only in cow cheeses. The residual fat content in the cheese extracts, instead, showed a smaller correlation with the antiproliferative effect on HeLa cells in ewe cheeses ($r = 0.663$) compared to cow samples ($r = 0.796$ or 0.790, with or without C2 cheese, respectively). This could be due to the presence of different small peptides and fats in the extracts according to milk origin, but also to cheese ripening, which plays an important role in the release and formation of new peptides and fats [52]: according to the presence of specific compounds, these could result in exerting different biological activities. Different inhibitory effects of water-soluble peptides (WSPs) extracts of Cheddar were observed by Rafiq, et al. [45] according to the milk's origin, with WSPs from buffalo Cheddar exerting a higher inhibitory effect on lung cancer cell line (H-1299) compared to WSPs from cow milk Cheddar. Furthermore, the same authors [45] observed different viability of the cancer cells during ripening. A positive correlation between ripeness and antiproliferative activity of various cheeses was also observed against human leukemia cellular models [43,49]. Finally, as for the N/protein, a higher antiproliferative effect on MDA-MB-231 cells was observed with increases in residual fat, but only in the crocin-rich extracts of cow cheeses ($r = 0.822$).

It is worthwhile highlighting that among cow cheeses, treatment with C2 extract, even if bearing a higher crocin content compared to other cow cheese extracts, did not show any significant effect on HeLa and MDA-MB-231 cell viability. On the other hand, cells treated with C3 cheese extract showed higher antiproliferative effects, confirming that chemical environment plays an important role in modulating crocin's antiproliferative effects. Furthermore, with equal crocin contents incubated on the cells compared to ewe cheeses, treatment with C2 sample showed no significant antiproliferative effect despite its higher residual N/protein and fat contents), supposing a dissimilar influence of the chemical environment according to milk origin.

Table 3. Pearson's correlation analysis of the crocin-rich extracts in ewe and cow cheeses.

	Protein N residue	Fat residue	HeLa Proliferation	MDA-MB-231 Proliferation
Ewe Cheeses				
Protein N residue	1			
Fat residue	0.844	1		
HeLa Proliferation	−0.961	−0.663	1	
MDA-MB-231 Proliferation	0.098	0.450	0.370	1
Cow Cheeses				
	Protein N residue	Fat residue	HeLa Proliferation	MDA-MB-231 Proliferation
Protein N residue	1			
Fat residue	0.970	1		
HeLa Proliferation	−0.675	−0.796	1	
MDA-MB-231 Proliferation	−0.688	−0.822	0.993	1

Abdullaev Jafarova, et al. [33] observed a different inhibitory behavior of the various saffron components on the colony formation of HeLa cellular lines, with *trans*-crocin 3 (*trans*-3-Gg) resulting in having the highest inhibitory activity, followed by picrocrocin in acidic form, picrocrocin and *trans*-crocin 4 (*trans*-4-GG). Furthermore, gentiobiosyl residues have been revealed to play an important role on the cytotoxic effect of saffron against tumor cells [53]. For these reasons, *trans*-3-Gg/*trans*-4-GG ratio was calculated in each crocin-rich extract (Table 2). Unfortunately, the major or minor presence of *trans*-3-Gg compared to *trans*-4-GG does not seem to explain the different behavior in the antiproliferative activity of the extracts: on the contrary, the cheese extracts with the lowest *trans*-3-Gg/*trans*-4-GG ratio showed the highest antiproliferative effects both on HeLa and MDA-MB-231 cells. This result could be related to a higher antiproliferative effect of *trans*-4-GG compared to *trans*-3-Gg, or to a greater contribution of the chemical environment to the antiproliferative effects of the cheese extracts compared to crocins. It was surprising to observe that a high correlation (r = 0.83) between antiproliferative activity and *trans*-4-GG percentage was observed only in cow cheese, whose r value rose up to 0.98 excluding C2 sample, which had double the crocins content compared to other cow cheeses extracts. This result would confirm the important role of the cheese matrix compounds in modulating the antiproliferative effect of crocins.

3. Materials and Methods

3.1. Samples and Chemicals

Different Italian cheeses, characterized by saffron addition during cheese making, were purchased in local markets or directly from the farmers. The samples evaluated in this study were as follows:

- three ewe milk cheeses (E1–E3) produced in three different farms and differing for ripening time and saffron addition.
- four cow milk cheeses (C1–C4), produced in the same farm and differing for indoor or local pasture feeding and for ripening time.

The main characteristics, as well as the way to which saffron was added during cheese making, are reported in Table 4. All dairy products were sampled and grated according to the reference method [54]. Three samples of the same cheese batch were analyzed, each one in triplicate.

The farm producing cow cheeses also provided saffron used in the cheese making.

Trans-crocetin di-(β-D-gentibiosyl) ester (*trans*-4-GG) (68% purity), safranal (purity 88%) and acetonitrile (for HPLC, ≥99.9%) were purchased from Sigma (Sigma–Aldrich Co., St. Louis, MO, USA), methanol (special grade, RS—for UHPLC-MS) from Carlo Erba

(Carlo Erba Reagents S.A.S, Val de Reuil, France), while water was purified through a ion exchange Milli-Q system (resistivity >18 mΩ; Millipore, Billerica, MA, USA).

Table 4. Main characteristics of the cow and ewe cheese samples.

Sample	Cheese Variety	Milk	Ripening	Saffron Addition
E1	Pecorino-type (ewe)	Pasteurized	30–40 days	Powder added during cutting of the curd
E2	Pecorino-type (ewe)	N.A. [1]	about 15 days	Stigmas added during cutting of the curd
E3	Pecorino-type (ewe)	Raw	10 months	Powder (previously suspended in milk) added after cutting of the curd
C1	Cow hard cheese	Raw partially skimmed	22 months	Powder added to milk, or to a mixture of whey and curd, before cutting of the curd
C2	Cow hard cheese	Raw partially skimmed	14 months	Powder added to milk, or to a mixture of whey and curd, before cutting of the curd
C3	Cow hard cheese	Raw partially skimmed	19 months	Powder added to milk, or to a mixture of whey and curd, before cutting of the curd
C4	Cow hard cheese	Raw partially skimmed	16 months	Powder added to milk, or to a mixture of whey and curd, before cutting of the curd

[1] N.A.: Not Available.

3.2. *Saffron Bioactive Molecules in Cheese*

Extraction and quantification of saffron bioactive molecules from cheese, namely crocins and picrocrocin, were carried out according to the method of Ritota, et al. [16]. Briefly 2.5 g of the grated cheese were extracted with 10 mL of a methanol:water (80:20, *v*/*v*) solution by stirring in the dark, at room temperature, for 1 h. The sample was then centrifuged (3500 rpm for 5 min at 4 °C) and the supernatant fraction was collected. Extraction was repeated twice more in the same way and the supernatant fractions were collected and filled to the mark (30 mL) with the same extraction solution. The extract was filtered (0.2 μm) before the UHPLC analysis. Chromatographic determination was carried out on a Nexera UHPLC system (Shimadzu Corporation, Kyoto, Japan) equipped with two LC-30AD pumps, UV–Vis SPD-M20A photodiode array (PDA) detector and SIL 30-AC autosampler, using a Synergi Fusion-RP (100 × 2.1, 2.5 μm, Phenomenex Inc., Torrance, CA, USA) column, with a SecurityGuard™ ULTRA cartridge for C18 UHPLC (2 × 2.1 mm, Phenomenex Inc., Torrance, CA, USA). LabSolution Version 5.42 SP5 (Shimadzu Corporation, Kyoto, Japan) software was used for equipment control, data acquisition and processing. Crocins were monitored at 440 nm, and the total crocins content was expressed as mg *trans*-4-GG/100 g cheese, by means of an external calibration curve for *trans*-4-GG (68% purity), and supposing the same response factor for all *cis* and *trans*-crocins. Picrocrocin was monitored at 250 nm.

Safranal content was detected in the same hydroalcoholic extract using the same chromatographic conditions proposed by Ritota, et al. [16], with PDA detector at 310 nm. Safranal content, expressed as mg/100 g cheese, was quantified by means of an external calibration curve (32–4235 μg/L). The recovery on a spiked sample was 72%, the instrumental Limit of Detection (LOD) and Limit of Quantification (LOQ) were equal to 0.006 and 0.014 mg/L, respectively (corresponding to 8 and 17 μg/100 g cheese), and the inter-day relative standard deviation (n = 3) was 0.4%.

3.3. *Cell Lines and Treatment*

HeLa, MDA-MB-231 and CaCo2 cells were obtained from the American Tissue Culture Collection (Manassas, VA, USA). The cells were grown in DMEM medium (Euroclone, Milan, Italy) supplemented with 10% Fetal Bovine Serum (FBS, Sigma–Aldrich Co., St. Louis, MO, USA), 100 U/mL penicillin and 100 μg/mL streptomycin (Pen/Strep, Euroclone, Milan, Italy), 2 mM glutamine (Euroclone, Milan, Italy) and 1% non-essential amino acid (Merck, Darmstadt, Germany). The cells were maintained at 37 °C in a humidified atmosphere of 5% CO_2/95% air. 40.000 cells were seeded onto 24-multiwell plates.

The hydroalcoholic extracts from cheeses, rich in crocins, were filtered (0.2 μm) and brought to dryness under nitrogen flow, then they were dissolved in DMSO and added to the culture medium for 24 h.

3.4. *In Vitro Viability Assay (MTT Assay)*

A total of 0.5 mg/mL of MTT reagent (Merck, Darmstadt, Germany) was added to each well and incubated for 3 h at 37 °C in the CO_2 incubator. The MTT solution was then discarded, and 100 μL of isopropanol was added. The plates were placed on a shaker to solubilize the formation of purple crystal formazan. The absorbance was measured using a microplate reader (Tecan 200 infinite pro) at a wavelength of 570 nm.

3.5. *Residual Protein Nitrogen and Fat in the Crocin-Rich Extracts*

The residual fat content in the hydroalcoholic crocin-rich extracts were determined according to the Bligh and Dyer [55] method. The residual protein nitrogen was determined according to the Kjeldahl method, using a multiplication factor of 6.38 [56], prior total evaporation of the hydroalcoholic crocin-rich extract and redissolution in ultrapure water.

3.6. *Statistics*

PAST 2.17 software [57] was used to perform statistical analysis. Normal data distribution and homogeneity of variance were tested using Shapiro–Wilk's test and Levene's test, respectively. One-way Analysis of Variance (ANOVA), followed by a Tukey's Post Hoc with Bonferroni correction, was applied to evaluate significant differences ($p < 0.05$) among the mean values. In case of non-normal distribution and non-homogeneity of variance, Friedman and Kruskal–Wallis tests, respectively, were applied to test significant differences ($p < 0.05$) among data.

4. Conclusions

In this study, the main bioactive compounds of saffron (crocins, picrocrocin and safranal) were evaluated for the first time in cow and ewe cheeses made with saffron by a single UHPLC analytical method. Crocins were detected in all cheese samples, with levels varying according to the different cheese making process. Picrocrocin was not detected in cheese probably due to its degradation during cheese making, while safranal was detected only in the ewe cheese with the highest crocins content.

With the antiproliferative activity of saffron being known, and due to the lack of evidence of the antiproliferative effects of crocins within a food matrix or in a complex mixture, for the first time hydroalcoholic extracts from cheese containing saffron were tested for their antiproliferative activities against three cancer cellular models, CaCo2 (colon cancer), HeLa (cervix carcinoma) and MDA-MB-231 cells. No effect was observed on CaCo2 cells, probably due to a great resistance of intestinal cell to external insults, while HeLa and MDA-MB-231 cells were sensitive to treatment with crocin-rich extracts from cheese. The antiproliferative effects of the hydroalcoholic crocin-rich extracts of the studied cheeses could be ascribed to saffron's bioactive compounds (mainly crocins), and to nitrogen and fat compounds present in the same extracts. However, a deeper characterization of the hydroalcoholic crocin-rich extracts could be helpful in better identifying the cheese compounds responsible, together with saffron crocins, for the antiproliferative effect in specific cell models (HeLa and MDA-MB-231, but not in CaCo2).

The chemical environment of the food matrix resulted in bearing a great influence in modulating antiproliferative effects of crocin: the crocin-rich extracts from cheese with both high residual N/protein and fat contents showed increased antiproliferative effects compared to pure crocin (*trans*-4-GG), probably due to milk origin and cheese ripening, both responsible for the presence of different type of fat compounds and small peptides.

The present study provides preliminary results concerning antiproliferative activity of extracts from cow and ewe cheeses made with saffron against cervix and breast carcinoma cell models. This could provide additional value to cheese made with saffron, both for farmers and consumers. Although these results are preliminary, additional studies are needed to investigate the antiproliferative effects of cheese extracts containing saffron in different cell lines and to research the molecular pathways involved. Finally, in order to evaluate the effect of bioactive compounds of saffron within food matrix, other studies should also be carried out on in vivo models (animals and/or in humans).

Author Contributions: Conceptualization, M.R., R.C. and P.M.; investigation, M.R. and R.C.; methodology, M.R. and R.C.; formal analysis, M.R. and R.C.; writing—original draft preparation, M.R.; writing—review and editing, M.R., R.C. and P.M. All authors have read and agreed to the published version of the manuscript.

Funding: This research received no external funding.

Institutional Review Board Statement: Not applicable.

Informed Consent Statement: Not applicable.

Data Availability Statement: The data presented in this study are available on request from the corresponding author.

Acknowledgments: This research was supported by the Italian Ministry of Agricultural, Food and Forestry Policies—BIOVITA Project.

Conflicts of Interest: The authors declare no conflict of interest.

Sample Availability: Samples of the compounds are not available from the authors.

References

1. Basker, D.; Negbi, M. Uses of saffron. *Econ. Bot.* **1983**, *37*, 228–236. [CrossRef]
2. Negbi, M. Saffron cultivation: Past, present and future prospects. In *Saffron*; CRC Press: Boca Raton, FL, USA, 1999; pp. 14–28.
3. Melnyk, J.P.; Wang, S.; Marcone, M.F. Chemical and biological properties of the world's most expensive spice: Saffron. *Food Res. Int.* **2010**, *43*, 1981–1989. [CrossRef]
4. Abu-Izneid, T.; Rauf, A.; Khalil, A.A.; Olatunde, A.; Khalid, A.; Alhumaydhi, F.A.; Aljohani, A.S.; Sahab Uddin, M.; Heydari, M.; Khayrullin, M. Nutritional and health beneficial properties of saffron (*Crocus sativus* L.): A comprehensive review. *Crit. Rev. Food Sci. Nutr.* **2020**. [CrossRef] [PubMed]
5. Akowuah, G.A.; Htar, T.T. Therapeutic properties of saffron and its chemical constituents. *J. Natl. Prod.* **2014**, *7*, 5–13.
6. Fernández, J.-A. Anticancer properties of saffron, Crocus sativus Linn. *Adv. Phytomed.* **2006**, *2*, 313–330.
7. Hoshyar, R.; Mollaei, H. A comprehensive review on anticancer mechanisms of the main carotenoid of saffron, crocin. *J. Pharm. Pharmacol.* **2017**, *69*, 1419–1427. [CrossRef] [PubMed]
8. Lambrianidou, A.; Koutsougianni, F.; Papapostolou, I.; Dimas, K. Recent advances on the anticancer properties of saffron (*Crocus sativus* L.) and its major constituents. *Molecules* **2021**, *26*, 86. [CrossRef]
9. Bathaie, S.Z.; Bolhasani, A.; Hoshyar, R.; Ranjbar, B.; Sabouni, F.; Moosavi-Movahedi, A.-A. Interaction of saffron carotenoids as anticancer compounds with ctDNA, Oligo (dG. dC) 15, and Oligo (dA. dT) 15. *DNA Cell Biol.* **2007**, *26*, 533–540. [CrossRef]
10. Festuccia, C.; Mancini, A.; Gravina, G.L.; Scarsella, L.; Llorens, S.; Alonso, G.L.; Tatone, C.; Di Cesare, E.; Jannini, E.A.; Lenzi, A. Antitumor effects of saffron-derived carotenoids in prostate cancer cell models. *BioMed Res. Int.* **2014**, *2014*, 135048. [CrossRef]
11. Aktypis, A.; Christodoulou, E.D.; Manolopoulou, E.; Georgala, A.; Daferera, D.; Polysiou, M. Fresh ovine cheese supplemented with saffron (*Crocus sativus* L.): Impact on microbiological, physicochemical, antioxidant, color and sensory characteristics during storage. *Small Rumin. Res.* **2018**, *167*, 32–38. [CrossRef]
12. Carpino, S.; Rapisarda, T.; Belvedere, G.; Licitra, G. Volatile fingerprint of Piacentinu cheese produced with different tools and type of saffron. *Small Rumin. Res.* **2008**, *79*, 16–21. [CrossRef]
13. Licón, C.; Carmona, M.; Molina, A.; Berruga, M. Chemical, microbiological, textural, color, and sensory characteristics of pressed ewe milk cheeses with saffron (*Crocus sativus* L.) during ripening. *J. Dairy Sci.* **2012**, *95*, 4263–4274. [CrossRef] [PubMed]
14. Librán, C.M.; Licón, C.C.; Serrano-Díaz, J.; Carmona, M.; Berruga, M.I. Safranal transference from ewe's milk to cheese and whey and antifungal properties of fortified whey. *Dairy Sci. Technol.* **2014**, *94*, 83–89. [CrossRef]
15. Licón, C.C.; Carmona, M.; Berruga, M.I. Volatile compounds in pressed ewes' milk cheese with saffron spice (*Crocus sativus* L.). *Int. J. Dairy Technol.* **2015**, *68*, 399–408. [CrossRef]
16. Ritota, M.; Mattera, M.; Costanzo, M.G.D.; Manzi, P. Evaluation of Crocins in Cheeses Made with Saffron by UHPLC. *J. Braz. Chem. Soc.* **2018**, *29*, 248–257. [CrossRef]
17. Aung, H.; Wang, C.; Ni, M.; Fishbein, A.; Mehendale, S.; Xie, J.; Shoyama, A.; Yuan, C. Crocin from Crocus sativus possesses significant anti-proliferation effects on human colorectal cancer cells. *Exp. Oncol.* **2007**, *29*, 175. [PubMed]
18. Chryssanthi, D.G.; Dedes, P.G.; Karamanos, N.K.; Cordopatis, P.; Lamari, F.N. Crocetin inhibits invasiveness of MDA-MB-231 breast cancer cells via downregulation of matrix metalloproteinases. *Planta Med.* **2011**, *77*, 146–151. [CrossRef]

19. Dhar, A.; Mehta, S.; Dhar, G.; Dhar, K.; Banerjee, S.; Van Veldhuizen, P.; Campbell, D.R.; Banerjee, S.K. Crocetin inhibits pancreatic cancer cell proliferation and tumor progression in a xenograft mouse model. *Mol. Cancer Ther.* **2009**, *8*, 315–323. [CrossRef]
20. Mousavi, S.H.; Moallem, S.A.; Mehri, S.; Shahsavand, S.; Nassirli, H.; Malaekeh-Nikouei, B. Improvement of cytotoxic and apoptogenic properties of crocin in cancer cell lines by its nanoliposomal form. *Pharm. Biol.* **2011**, *49*, 1039–1045. [CrossRef]
21. Bajbouj, K.; Schulze-Luehrmann, J.; Diermeier, S.; Amin, A.; Schneider-Stock, R. The anticancer effect of saffron in two p53 isogenic colorectal cancer cell lines. *BMC Complement. Altern. Med.* **2012**, *12*, 69. [CrossRef] [PubMed]
22. D'Alessandro, A.M.; Mancini, A.; Lizzi, A.R.; De Simone, A.; Marroccella, C.E.; Gravina, G.L.; Tatone, C.; Festuccia, C. Crocus sativus stigma extract and its major constituent crocin possess significant antiproliferative properties against human prostate cancer. *Nutr. Cancer* **2013**, *65*, 930–942. [CrossRef] [PubMed]
23. Samarghandian, S.; Boskabady, M.H.; Davoodi, S. Use of in vitro assays to assess the potential antiproliferative and cytotoxic effects of saffron (*Crocus sativus* L.) in human lung cancer cell line. *Pharm. Mag.* **2010**, *6*, 309. [CrossRef] [PubMed]
24. Caballero-Ortega, H.; Pereda-Miranda, R.; Abdullaev, F.I. HPLC quantification of major active components from 11 different saffron (*Crocus sativus* L.) sources. *Food Chem.* **2007**, *100*, 1126–1131. [CrossRef]
25. Masi, E.; Taiti, C.; Heimler, D.; Vignolini, P.; Romani, A.; Mancuso, S. PTR-TOF-MS and HPLC analysis in the characterization of saffron (*Crocus sativus* L.) from Italy and Iran. *Food Chem.* **2016**, *192*, 75–81. [CrossRef] [PubMed]
26. Tarantilis, P.A.; Polissiou, M.G. Chemical analysis and antitumor activity of natural and semi-natural carotenoids of saffron. In Proceedings of the I International Symposium on Saffron Biology and Biotechnology 650, Albacete, Spain, 22–25 October 2003; pp. 447–461.
27. Chavez, K.J.; Garimella, S.V.; Lipkowitz, S. Triple negative breast cancer cell lines: One tool in the search for better treatment of triple negative breast cancer. *Breast Dis.* **2010**, *32*, 35–48. [CrossRef] [PubMed]
28. Liu, H.; Zang, C.; Fenner, M.; Possinger, K.; Elstner, E. PPARγ ligands and ATRA inhibit the invasion of human breast cancer cells in vitro. *Breast Cancer Res. Treat.* **2003**, *79*, 63–74. [CrossRef] [PubMed]
29. Hire, R.R.; Srivastava, S.; Davis, M.B.; Kumar Konreddy, A.; Panda, D. Antiproliferative activity of crocin involves targeting of microtubules in breast cancer cells. *Sci. Rep.* **2017**, *7*, 44984. [CrossRef]
30. Calvert, T.; Aidoo, K.; Candlish, A.; Fuat, A.M. Comparison of in vitro cytotoxicity of Fusarium mycotoxins, deoxynivalenol, T-2 toxin and zearalenone on selected human epithelial cell lines. *Mycopathologia* **2005**, *159*, 413–419. [CrossRef] [PubMed]
31. William-Faltaos, S.; Rouillard, D.; Lechat, P.; Bastian, G. Cell cycle arrest and apoptosis induced by oxaliplatin (L-OHP) on four human cancer cell lines. *Anticancer Res.* **2006**, *26*, 2093–2099. [PubMed]
32. Zhou, C.; Tabb, M.M.; Sadatrafiei, A.; Grün, F.; Blumberg, B. Tocotrienols activate the steroid and xenobiotic receptor, SXR, and selectively regulate expression of its target genes. *Drug Metabol. Dispos.* **2004**, *32*, 1075–1082. [CrossRef] [PubMed]
33. Abdullaev Jafarova, F.; Caballero Ortega, H.; Riverón Negrete, L.; Pereda Miranda, R.; Rivera Luna, R.; Hernández, J.M.; Pérez López, I.; Espinosa Aguirre, J.J. Evaluación in vitro del potencial quimiopreventivo del azafrán. *Rev. Investig. Clín.* **2002**, *54*, 430–436.
34. Gezici, S. Comparative anticancer activity analysis of saffron extracts and a principle component, crocetin for prevention and treatment of human malignancies. *J. Food Sci. Technol.* **2019**, *56*, 5435–5443. [CrossRef]
35. Givens, D.I. Saturated fats, dairy foods and health: A curious paradox? *Nutr. Bull.* **2017**, *42*, 274–282. [CrossRef]
36. Lecerf, J.-M.; Legrand, P. Les effets des nutriments dépendent-ils des aliments qui les portent? L'effet matrice. *Cah. Nutr. Diététique* **2015**, *50*, 158–164. [CrossRef]
37. Zou, L.; Liu, W.; Liu, C.; Xiao, H.; McClements, D.J. Utilizing food matrix effects to enhance nutraceutical bioavailability: Increase of curcumin bioaccessibility using excipient emulsions. *J. Agric. Food Chem.* **2015**, *63*, 2052–2062. [CrossRef]
38. Capuano, E.; Oliviero, T.; van Boekel, M.A. Modeling food matrix effects on chemical reactivity: Challenges and perspectives. *Crit. Rev. Food Sci. Nutr.* **2018**, *58*, 2814–2828. [CrossRef]
39. Fardet, A. Food health potential is primarily due to its matrix structure, then nutrient composition: A new paradigm for food classification according to technological processes applied. *J. Nutr. Health Food Eng.* **2014**, *1*, 208–209. [CrossRef]
40. Mohanty, D.; Mohapatra, S.; Misra, S.; Sahu, P. Milk derived bioactive peptides and their impact on human health–A review. *Saudi J. Biol. Sci.* **2016**, *23*, 577–583. [CrossRef] [PubMed]
41. Meisel, H.; FitzGerald, R.J. Biofunctional peptides from milk proteins: Mineral binding and cytomodulatory effects. *Curr. Pharm. Design* **2003**, *9*, 1289–1296.
42. López-Expósito, I.; Miralles, B.; Amigo, L.; Hernández-Ledesma, B. Health effects of cheese components with a focus on bioactive peptides. In *Fermented Foods in Health and Disease Prevention*; Elsevier: Amsterdam, The Netherlands, 2017; pp. 239–273.
43. Yasuda, S.; Ohkura, N.; Suzuki, K.; Yamasaki, M.; Nishiyama, K.; Kobayashi, H.; Hoshi, Y.; Kadooka, Y.; Igoshi, K. Effects of highly ripened cheeses on HL-60 human leukemia cells: Antiproliferative activity and induction of apoptotic DNA damage. *J. Dairy Sci.* **2010**, *93*, 1393–1400. [CrossRef]
44. Rafiq, S.; Gulzar, N.; Huma, N.; Hussain, I.; Murtaza, M.S. Evaluation of anti-proliferative activity of Cheddar cheeses using colon adenocarcinoma (HCT-116) cell line. *Int. J. Dairy Technol.* **2020**, *73*, 255–260. [CrossRef]
45. Rafiq, S.; Huma, N.; Gulzar, N.; Murtaza, M.A.; Hussain, I. Effect of cheddar cheese peptide extracts on growth inhibition, cell cycle arrest and apoptosis induction in human lung cancer (H-1299) cell line. *Int. J. Dairy Technol.* **2018**, *71*, 975–980. [CrossRef]
46. Rafiq, S.; Huma, N.; Rakariyatham, K.; Hussain, I.; Gulzar, N.; Hayat, I. Anti-inflammatory and anticancer activities of water-soluble peptide extracts of buffalo and cow milk Cheddar cheeses. *Int. J. Dairy Technol.* **2018**, *71*, 432–438. [CrossRef]

47. De Simone, C.; Picariello, G.; Mamone, G.; Stiuso, P.; Dicitore, A.; Vanacore, D.; Chianese, L.; Addeo, F.; Ferranti, P. Characterisation and cytomodulatory properties of peptides from Mozzarella di Bufala Campana cheese whey. *J. Pept. Sci.* **2009**, *15*, 251–258. [CrossRef] [PubMed]
48. Rodríguez-Alcalá, L.M.; Castro-Gómez, M.P.; Pimentel, L.L.; Fontecha, J. Milk fat components with potential anticancer activity—A review. *Biosci. Rep.* **2017**, *37*, BSR20170705. [CrossRef] [PubMed]
49. Yasuda, S.; Kuwata, H.; Kawamoto, K.; Shirakawa, J.; Atobe, S.; Hoshi, Y.; Yamasaki, M.; Nishiyama, K.; Tachibana, H.; Yamada, K. Effect of highly lipolyzed goat cheese on HL-60 human leukemia cells: Antiproliferative activity and induction of apoptotic DNA damage. *J. Dairy Sci.* **2012**, *95*, 2248–2260. [CrossRef]
50. Snow, D.R.; Jimenez-Flores, R.; Ward, R.E.; Cambell, J.; Young, M.J.; Nemere, I.; Hintze, K.J. Dietary milk fat globule membrane reduces the incidence of aberrant crypt foci in Fischer-344 rats. *J. Agric. Food Chem.* **2010**, *58*, 2157–2163. [CrossRef] [PubMed]
51. Zanabria, R.; Tellez, A.M.; Griffiths, M.; Corredig, M. Milk fat globule membrane isolate induces apoptosis in HT-29 human colon cancer cells. *Food Funct.* **2013**, *4*, 222–230. [CrossRef]
52. Santiago-López, L.; Aguilar-Toalá, J.E.; Hernández-Mendoza, A.; Vallejo-Cordoba, B.; Liceaga, A.M.; González-Córdova, A.F. Invited review: Bioactive compounds produced during cheese ripening and health effects associated with aged cheese consumption. *J. Dairy Sci.* **2018**, *101*, 3742–3757. [CrossRef]
53. Bathaie, S.Z.; Mousavi, S.Z. New applications and mechanisms of action of saffron and its important ingredients. *Crit. Rev. Food Sci. Nutr.* **2010**, *50*, 761–786. [CrossRef]
54. *ISO 707/IDF 50*; Milk and Milk Products—Guidance on Sampling. ISO: Genève, Switzerland, 2008.
55. Bligh, E.G.; Dyer, W.J. A rapid method of total lipid extraction and purification. *Can. J. Biochem. Physiol.* **1959**, *37*, 911–917. [CrossRef] [PubMed]
56. GU (Gazzetta Ufficiale della Repubblica Italiana). *Metodi Ufficiali di Analisi per i Formaggi*; GU: Rome, Italy, 1986; Volume Serie generale n. 229, pp. 1–40.
57. Hammer, Ø.; Harper, D.; Ryan, P. PAST, v. 2.17c. *Palaeontol. Electron.* **2001**, *4*, 1–229.

molecules

MDPI

Article

Effect of Vitamin C Source on Its Stability during Storage and the Properties of Milk Fermented by *Lactobacillus rhamnosus*

Agata Znamirowska, Katarzyna Szajnar * and Małgorzata Pawlos

Department of Dairy Technology, Institute of Food Technology and Nutrition, College of Natural Sciences, University of Rzeszow, Cwiklińskiej 2D, 35-601 Rzeszów, Poland; aznamirowska@ur.edu.pl (A.Z.); mpawlos@ur.edu.pl (M.P.)

* Correspondence: katarzyna.szajnar@gmail.com

Abstract: The enrichment of commonly consumed foods with bioactive components might be helpful in promoting health and reducing the risk of disease, so the enrichment of probiotic fermented milk with vitamin C can be considered appropriate. The effect of vitamin C addition depends on the source of origin (rosehip, acerola and ascorbic acid in powder form) on the growth and survival of *Lactobacillus rhamnosus* and the quality of fermented milk on the 1st and 21st day of storage was analyzed. The pH, total acidity, vitamin C, syneresis, color, texture profile and numbers of bacterial cells in fermented milk were determined. The organoleptic evaluation was also performed. The degradation of vitamin C in milk was shown to depend on its source. The lowest reduction of vitamin C was determined in milk with rosehip. The least stable was vitamin C naturally found in control milk. The addition of rosehip and acerola decreased syneresis and lightness of milk color, increasing the yellow and red color proportion. In contrast, milk with ascorbic acid was the lightest during the whole experimental period and was characterized by a very soft gel. The growth of *Lactobacillus rhamnosus* during fermentation was most positively affected by the addition of rosehip. However, the best survival of *Lactobacillus rhamnosus* was demonstrated in milk with acerola. On the 21st day of storage, the number of *L. rhamnosus* cells in the control milk and the milk with vitamin C was >8 log cfu g^{-1}, so these milks met the criterion of therapeutic minimum. According to the assessors, the taste and odor contributed by the addition of rosehip was the most intense of all the vitamin C sources used in the study.

Keywords: fermented milk; vitamin C; *Lactobacillus rhamnosus*; probiotic bacteria

1. Introduction

The consumption of probiotic foods is steadily increasing, and consumers are choosing these products especially for their taste and beneficial health aspects. It is assumed that probiotic bacteria have the potential to provide health benefits that include intestinal microbial balance and mineral absorption, prevention of constipation, reduction of cholesterol levels in serum, and improvements in lactose intolerance and blood pressure [1–3].

Lactobacillus rhamnosus (LGG) with probiotic properties is a promising natural alternative to commercial additives in the food industry [4]. This bacteria exhibits most of the characteristics generally desirable for a good probiotic strain, including the ability to survive passage through the human gastrointestinal tract after ingestion and the capacity to temporarily colonize the ileum and colon. The beneficial health properties of LGG depend partially on its prolonged residence in the gastrointestinal tract and strong adhesion ability. Therefore, LGG is an effective starter for fermented milk production, while fermented milk with LGG has various probiotic functions for adults and children. Moreover, this bacilli has the benefits of being resistant to low pH levels and has been reported to show excellent viability [5]. Furthermore, it is a potential candidate for the synthesis of acetoin and diacetyl, which are butter-like flavor compounds widely used in the dairy industry [6].

However, it is crucial to ensure an adequate number of viable bacterial cells, which is the "therapeutic minimum" that should be regularly consumed to achieve the "probiotic" effect for the consumer. According to the International Dairy Federation Recommendation, probiotic products should contain at least seven log CFU g^{-1} of lactic acid bacteria. Therefore, the survival of these bacteria during the shelf-life and until consumption is an important aspect. Furthermore, criteria for the selection of probiotic strains include safety as well as technological, sensory and functional properties of the strains [7,8].

The enrichment of commonly consumed foods with bioactive components may help promote health and reduce disease risk, so the enrichment of probiotic fermented milk with vitamin C may be considered appropriate.

Due to the very good solubility of vitamin C and active transport, it is absorbed by the body in about 70–80% (from a dose of 180 mg in non-smokers) [9,10]. The main organs where the absorption process occurs are the duodenum and the proximal part of the small intestine [10,11]. The efficiency of this process strongly depends on the condition of the body. It might be impaired by vomiting, lack of appetite, digestion and absorption disorders, intestinal dysfunction, smoking and use of some drugs (e.g., aspirin) [9,11].

Vitamin C is the most well-known antioxidant. With its antioxidant properties, this vitamin provides a protective role in cardiovascular disease. Zhang et al. [12], in a study conducted on cigarette smokers, demonstrated that ascorbic acid in conjunction with other antioxidants (including vitamin E) inhibited elevated markers of lipid peroxidation induced by smoking in smokers. Antioxidants, including vitamin C, were confirmed to have a protective role in coronary heart disease and cardiovascular disease. Based on its antioxidant capacity, ascorbic acid protects body cells from oxidative stress [12].

Acerola (*Malpighia emarginata* DC.) is one of the richest natural sources of ascorbic acid in the world. It is grown in areas from southern Texas, Mexico, and Central America to northern South America and the Caribbean. It is recently introduced to subtropical areas around the world, including India. The fruit is consumed as fresh juice and as cherry consumed daily by many native tribes, or by animals [13]. Acerola is a good source of vitamin C, which also contains amino acids, phenolic compounds, and carotenoids, classifying it as a nutraceutical [14–16]. The vitamin C content of acerola fruit varies depending on ripening stages, cultivar, growing location, and environmental factors [15,17,18]. Vitamin C content is high in unripe fruit, about 1.9% of juice and decreases during ripening to about 0.97% of juice in ripe fruit [17].

Rosehips are wild pseudo fruits derived from plants in the genus *Rosa*, widely cultivated in Europe, the Middle East, Asia, and North America. The most abundant and most studied species in Europe is *Rosa canina*, which is a native shrub. This fruit is usually consumed as tea, jelly, jam and beverages, and is now also used as an ingredient in probiotic drinks, yoghurts and soups [19]. Rose hips are also rich in vitamin C. The content of L-ascorbic acid in these fruits ranges from 0.5 to 2.0%. Moreover, rose hips contain vitamins, e.g., B1, B2, PP, A, E, K, carotenoids, and polyphenolic compounds [20].

It was shown that the stability of ascorbic acid in rosehip fruit is higher in the fruit matrix than in the extracts. Furthermore, flavonoids from rosehips could prevent the oxidation of ascorbic acid [21,22].

The bioavailability of water-soluble vitamins in the gastrointestinal tract, including the oral cavity, stomach, and small intestine, could vary depending on pH, temperature, binding to polypeptides and polysaccharides and the presence of metal ions and digestive enzyme inhibitors [23].

Depending on the additional compounds present in acerola and rosehip and the resulting stability of vitamin C in beverages, a study was conducted on three different sources of vitamin C: rosehip, acerola and ascorbic acid. Therefore, the effect of vitamin C addition, in dependence on the source, on the growth and survival of *Lactobacillus rhamnosus* and the quality of fermented milk on the 1st and 21st day of storage was analyzed.

2. Results and Discussion

2.1. Physicochemical Properties of Fermented Milk

The pH Value and Total Acidity

The milk was enriched with vitamin C at a dose of approximately 30 ± 3 mg 100 g^{-1} in the form of rosehip (DR), acerola (AC) and ascorbic acid (VC). As expected, the pH value of DR, AC and VC milk with vitamin C addition was significantly lower before fermentation compared to the control milk (Table 1). Before fermentation, the addition of rosehip to DR milk decreased the pH value the most, whereas after fermentation, the pH was the highest in DR milk with rosehip and VC with ascorbic acid, which was also associated with the lowest lactic acid content in this milk. On the 1st and 21st day of storage, the highest amount of lactic acid was determined in AC milk with acerola. Lactic acid content did not change significantly during refrigerated storage of milk, which is also confirmed by a two-way analysis of variance (Table 2).

Table 1. Vitamin C, color, syneresis, pH and total acidity of fermented milk.

Properties		Storage Time	K [1]	DR [2]	AC [3]	VC [4]
pH		0	6.65 ^{cB} ± 0.03	5.33 ^{aC} ± 0.11	6.52 ^{bC} ± 0.06	6.26 ^{dC} ± 0.02
		1	4.65 ^{bA} ± 0.30	4.86 ^{cB} ± 0.02	4.38 ^{aB} ± 0.01	4.76 ^{bB} ± 0.05
		21	4.39 ^{aA} ± 0.09	4.70 ^{cA} ± 0.02	4.30 ^{aA} ± 0.01	4.42 ^{bA} ± 0.01
Total acidity, g lactic acid L^{-1}		1	0.72 ^{bA} ± 0.02	0.69 ^{bA} ± 0.01	0.83 ^{cA} ± 0.05	0.65 ^{aA} ± 0.04
		21	0.74 ^{aA} ± 0.14	0.70 ^{aA} ± 0.01	0.75 ^{aA} ± 0.07	0.72 ^{aA} ± 0.04
Vitamin C, mg 100 g^{-1}		0	10.58 ^{aC} ± 0.36	43.33 ^{bC} ± 0.20	43.06 ^{bC} ± 0.30	43.72 ^{bB} ± 0.30
		1	9.52 ^{aB} ± 0.26	42.30 ^{bB} ± 0.20	42.00 ^{bB} ± 0.10	43.06 ^{cB} ± 0.39
		21	5.10 ^{aA} ± 0.51	40.72 ^{bA} ± 0.11	40.30 ^{bA} ± 0.21	40.22 ^{bA} ± 0.10
Syneresis, %		1	64.23 ^{bA} ± 1.18	61.93 ^{aA} ± 1.05	72.53 ^{cB} ± 0.59	61.37 ^{aA} ± 1.29
		21	65.86 ^{aA} ± 0.66	60.20 ^{cA} ± 0.38	64.05 ^{bA} ± 0.34	63.19 ^{bB} ± 0.96
Color	L*	1	94.96 ^{cA} ± 1.85	58.02 ^{aB} ± 1.66	86.95 ^{bB} ± 1.21	95.04 ^{cA} ± 1.46
		21	90.60 ^{aA} ± 4.26	51.42 ^{aA} ± 2.52	82.38 ^{bA} ± 0.22	94.20 ^{dA} ± 0.42
	a*	1	−1.81 ^{aA} ± 0.29	3.88 ^{cA} ± 0.30	1.54 ^{bA} ± 0.25	−1.55 ^{aB} ± 0.18
		21	−1.89 ^{aA} ± 0.35	6.37 ^{dB} ± 0.27	1.69 ^{cA} ± 0.33	−1.32 ^{bA} ± 0.02
	b*	1	8.54 ^{aA} ± 0.97	17.74 ^{cA} ± 1.21	9.06 ^{bA} ± 0.55	6.89 ^{aA} ± 0.47
		21	8.12 ^{bA} ± 0.21	19.24 ^{dA} ± 0.74	10.20 ^{cA} ± 0.89	6.84 ^{aA} ± 0.12
	C	1	8.76 ^{aA} ± 1.98	18.16 ^{bA} ± 1.21	9.19 ^{aA} ± 1.57	7.06 ^{aA} ± 0.47
		21	8.39 ^{bA} ± 0.24	20.56 ^{dB} ± 0.04	10.35 ^{cA} ± 0.91	6.96 ^{aA} ± 0.11
	h^0	1	102.30 ^{cA} ± 1.34	77.67 ^{aB} ± 0.31	99.72 ^{bA} ± 0.42	102.72 ^{cA} ± 1.52
		21	103.37 ^{cA} ± 0.28	72.07 ^{aA} ± 0.99	99.42 ^{bA} ± 0.88	100.92 ^{bA} ± 0.29

[1]—control milk; [2]—fermented milk with rosehip; [3]—fermented milk with acerola; [4]—fermented milk with L(+)ascorbic acid; 0—before fermentation; 1—after 1st day of storage; 21—after 21st day of storage; Values are mean ± SD; n = 15 for each group; A,B—mean values in columns denoted by different letters differ statistically significantly ($p < 0.05$) depending on the storage time; a–c—mean values in lines denoted by different letters differ statistically significantly ($p < 0.05$) depending on the different forms of vitamin C. Color: L* as brightness; a* color from red (+) to green (-); b* as the colors from yellow (+) to blue (-); C as the purity and intensity of the color, h^0 as the shade of the color.

In Linares et al. [24] study, L. *rhamnosus* was suspected of assimilating L-ascorbate, and in vitro experiments were performed to evaluate the putative catabolism. The results showed a global increase in the maximum absorbance obtained in the presence of D-glucose or L-ascorbate. There was a correlation with the presence of carbohydrates, as no increase in absorbance was observed in the negative control (no carbohydrate). However, control abiotic tests (performed without cell suspension) showed that this phenomenon was partly due to the yellow coloration of the substrate observed in the presence of L-ascorbate. In the study of Linares et al. [24], the growth of L. *rhamnosus* GG was supported by L-ascorbate because it was the only amino acid compound that could be used as a carbon source. L-ascorbate might serve as a reducing agent or a cofactor to promote utilising other carbon sources present in the medium (for example, amino acids). The main hypothesis explaining the presence of 12C-lactate is the catabolism of glucogenic amino acids. The biotransformation of 12C-amino acids could generate 12C-pyruvate and consequently 12C-lactate through lactate dehydrogenase activity. This pathway could also lead to the production of 12C-acetate. According to the metabolism of 13C-ascorbate, 12C-lactate might be formed only by amino acid catabolism. After 67 h of growth, the experimental results clearly showed that when bacteria assimilated amino acids, the kinetics of lactate production showed a much higher increase than the kinetics of acetate, which increased more or less consistently. Amino acid degradation generates more lactate than acetate (about three times more), while the opposite is true for ascorbate-only degradation. Acerola is rich in vitamin C, pectin, pectolytic enzymes, carotenoids and fiber, and vitamins and bioactive substances such as thiamine, riboflavin, niacin, proteins and mineral salts, mainly iron, calcium and phosphorus. The amino acid composition of some varieties of acerola (Vietnam) was notable, with an exceptionally high concentration of proline (858 mg kg^{-1}), accounting for almost half of the total amino acid content (1733 mg kg^{-1}). These fruits also contained high levels of phenylpropanoids, representing 30% of the total polyphenol content [15]. This fact might be explained by the high total acidity expressed as lactic acid in AC milk compared to DR and VC milk.

2.2. The Vitamin C

Before fermentation, 10.5 mg 100 g^{-1} of vitamin C was determined in control milk, whereas in milk with added vitamin C from different sources, the vitamin C content was 43 mg 100 g^{-1}. After fermentation, the vitamin C content decreased significantly in all milk groups, even in control milk, compared to the amount of vitamin C before fermentation. Fermentation by *Lactobacillus rhamnosus* of control milk K and rosehip DR, and acerola AC decreased the vitamin C content by about 1 mg 100 g^{-1}. In comparison, in VC milk with ascorbic acid, it only reduced by 0.7 mg 100 g^{-1}.

Our study did not show high vitamin C degradation, and there are reports that the acidic environment stabilizes vitamin C. According to Pérez-Vicente et al. [25], loss of vitamin C in pomegranate juice occurred due to pH changes. Yuan and Chen [26] studied the distribution of vitamin C in aqueous solutions. They demonstrated that low pH conditions promoted the formation of furfural, 2-furoesic acid and 3-hydroxy-2-pyrone. Some authors [27] indicate that a specific increase in the vitamin C content of fermented dairy products might be related to its synthesis by selected lactic acid bacteria. Ascorbic acid as an antioxidant reduces the number of substances formed during microbial metabolism [28]. In this study, extending the storage time of fermented milk to 21 days resulted in a continued decrease in vitamin C content from 4.4 mg 100 g^{-1} in control milk to 1.5 mg 100 g^{-1} in DR milk compared to the amount of vitamin C on the first day of storage. This implies that the least stable vitamin C was in the control milk, where its loss was 51.8% compared to its content before milk fermentation.

Differences in vitamin C content according to the source of origin were also determined. In DR milk with rosehip, the reduction of vitamin C content during storage was the lowest (5.9%). In VC milk with ascorbic acid, the degree of vitamin C degradation was the highest and amounted to 8.0% concerning its content before milk fermentation.

The least reduction of vitamin C in DR milk during storage might be related to the rosehip specific antioxidants. Phenolic compounds, including tannins, flavonoids, phenolic acids and anthocyanins, are an essential group of biologically active components in rosehip fruits [29]. Phenols are well known for their antioxidant properties, and several studies investigate the content and composition of polyphenols in different rose species. However, various quantitative and qualitative reports of the phenolic profile of roses are found in the literature. The main flavonoids are methyl gallate, catechin [30,31], epicatechin [31–33], rutin, eriocitrin, quercetin, apigenin-7-*O*-glucoside, kaempferol, [30, 31], quercetin and quinic acid [32]. Tumbas et al. [33] and Hosni et al. [34] identified quercetin and ellagic acid. In contrast, Türkben et al. [35] and Olsson et al. [36] found quercetin and catechin to be the essential phenolic components in rose fruits with the absence of ellagic acid and kaempferol. Demir et al. [30] and Elmastas et al. [31] found phenolic acids in rosehip fruits, including gallic acid, 4-hydroxybenzoic acid, caftaric acid, 2,5-dihydroxybenzoic acid, chlorogenic acid, t-caffeic acid, p-coumaric acid and ferulic acid. Nadpal et al. [32] reported protocatechuic acid except for the previously mentioned compounds.

Table 2. Analysis of variance (ANOVA) *p*-values on the effects of storage time and the source of vitamin C addition on vitamin C, pH, total acidity, syneresis, color parameters: L*, a*, b*, C, h^0, hardness, adhesiveness, stringiness length, cohesiveness, springiness, preference, consistency, milky-creamy taste, sour taste, taste of additives, sweet taste, off taste, sour odor, odor of additives, off odor, *L. rhamnosus* of fermented milk.

Properties	Storage Time *p*-Values	Source of Vitamin C *p*-Values	Storage Time × Source of Vitamin C *p*-Values
Vitamin C	0.0018↑	0.1010 n.s.	0.0510 n.s.
pH	0.0003↑	0.0003↑	0.0138↑
Total acidity	0.9937 n.s.	0.0139↑	0.5691 n.s.
Syneresis	0.0073↑	0.0262↑	0.0065↑
L*	0.0907↑	0.0000↑	0.0091↑
a*	0.0000↑	0.0000↑	0.0000↑
b*	0.0907 n.s.	0.0000↑	0.0863 n.s.
C	0.0719 n.s.	0.0000↑	0.0482
h^0	0.0503 n.s.	0.0000↑	0.0512 n.s.
Hardness	0.7351 n.s.	0.0000↑	0.8059 n.s.
Adhesiveness	0.6535 n.s.	0.1805 n.s.	0.7476↑
Cohesiveness	0.3739 n.s.	0.0860 n.s.	0.9642 n.s.
Springiness	0.4008 n.s.	0.0514 n.s.	0.7199 n.s.
Consistency	0.0002↑	0.0006↑	0.0003↑
Milky-creamy taste	0.0004↑	0.0126↑	0.0261↑
Sour taste	0.0003↑	0.1947 n.s.	0.1588 n.s.
Taste of additives	0.0824 n.s.	0.0340↑	0.4515 n.s.
Sweet taste	0.0321↑	0.0451↑	0.0472↑
Off-taste	0.8451 n.s.	0.7142 n.s.	0.9411 n.s.
Fermentation odor	0.0295↑	0.0140↑	0.0092↑
Odor of additives	0.7419 n.s	0.0121↑	0.8134 n.s.
Off-odor	0.6912 n.s.	0.4120 n.s.	0.0529 n.s.
L. rhamnosus	0.0228↑	0.0362↑	0.0014↑

Storage time × Source of vitamin C = interaction; ↑ indicates significant effect $p \leq 0.05$; n.s.: no significant effect.

Ilic and Ashoor [37] indicate that raspberry-infused yoghurt contained more reductants which decreased the rate of vitamin C degradation during storage.

The loss of vitamin C content in milk and dairy products during storage is affected by light, oxygen, iron, copper; moreover, it depends on the type of packaging [38]. FAO/WHO recommends 45–70 mg/day RNI (Recommended Nutrient Intakes) for vitamin C for an adult. Considering the vitamin C content in milk in our study, consumption of 100–150 mL of DR, VC and AC milk would meet the daily requirement for vitamin C [39].

2.3. Syneresis

The results of syneresis of milk fermented by *L. rhamnosus* are shown in Table 1. Both control milk and milk with added vitamin C from different sources (DR, AC, VC) characterized high syneresis. After fermentation, the highest syneresis was determined in AC milk, where the pH value was also the lowest and the lactic acid content the highest. Many authors explain that the ability to hold water in an acid gel depends on the acidity of the milk [40–42].

The lowest level of syneresis after fermentation was found in VC milk (61.37%) and DR milk (61.93%), in which the highest pH value was determined. An enhancement of syneresis with increasing storage time was found in the control milk and the milk with ascorbic acid (VC). However, in milk with rosehip (DR) and acerola (AC), an increase in storage time from 1 to 21 days resulted in a decrease in separated whey. The two-way analysis of variance indicated that the syneresis of fermented milk depended on the storage time, source of vitamin C, and these factors' interaction (Table 2). According to Sun et al. [43], milk fermented by *L. rhamnosus* showed a water retention capacity ranging from 34.7–39.3% calculated as a percentage of remaining sediment after centrifugation. In our study, syneresis was given as a percentage of whey separated, hence relating the results to the study of Sun et al. [43] could be assumed to be similar. Furthermore, these authors showed that nutrient enrichment of milk contributes to water retention capacity. Concerning our study, the addition of rosehip and ascorbic acid improved the water retention capacity of milk fermented by *L. rhamnosus*.

2.4. Color

The color components of milk fermented by *L. rhamnosus* depended on the source of added vitamin C. The lightness of L* was most intensely reduced by the addition of rosehip due to carotenoids, polyphenolic compounds such as tannins, flavonoids and anthocyanins in rosehip [44] (Table 1). Reduction in color lightness was also caused by the addition of acerola, which also contains phenolic compounds including anthocyanins and flavonoids and carotenoids [45]. In this study, the addition of ascorbic acid did not reduce color lightness on the first day of storage, and after 21 days, the milk with ascorbic acid was the lightest. Increasing the storage time to 21 days resulted in lower L* values in K, DR and AC milk. Similar to the control sample, VC milk with ascorbic acid showed a proportion of blue color and a smaller proportion of red color than DR and AC milk. The ratio of red and yellow increased most intensively with the addition of rosehip to the milk.

The redness (a*) and yellowness (b*) of DR and AC fermented milk increased during storage, contributing to a decrease in color lightness (L*). The obtained results indicate that the addition of rosehip and acerola to fermented milk might affect the color properties of milk. However, the addition of ascorbic acid differentiated the color components of fermented milk to the minimum degree during storage. Anthocyanins found in rosehip and acerola indicate a wide range of colors depending on pH. Due to the reduction in milk pH during fermentation of DR and AC milk by *L. rhamnosus*, the a* value of milk and DR and AC increased with a more extended storage period. The increase in b* value and decrease in L* lightness of DR and AC milk during storage may be related to the formation of brown polymers due to the degradation of phenolic compounds [46]. Analysis of variance ANOVA indicates that the C and h^0 parameters of fermented milk are only significantly affected by the source of vitamin C, whereas storage time and interaction of these two factors were not significant.

2.5. Parameters of Texture

Hardness is the compressive force of food or sample between the tongue and palate, as the attribute that most influences consumer choice products such as yoghurt [47,48]. Adhesiveness is a parameter that shows how much the product adheres to the probe (mouth) and elastic force; it is necessary to grind this semi-solid material until it is swallowed [49,50].

The results of the TPA test in Table 3 determine the texture properties (hardness, adhesiveness, cohesiveness, springiness) of all groups of milk fermented by *L. rhamnosus*. DR and AC milk showed higher hardness and adhesiveness values compared to control milk.

Table 3. Parameters of texture profile of milk fermented by *Lactobacillus rhamnosus* depending on the source of vitamin C.

Properties	Storage Time	K [1]	DR [2]	AC [3]	VC [4]
Hardness, N	1	0.91 ^{aA} ± 0.52	2.29 ^{bA} ± 0.18	1.37 ^{bA} ± 0.20	0.78 ^{aA} ± 0.05
	21	0.92 ^{aA} ± 0.45	1.93 ^{bA} ± 0.27	1.51 ^{bA} ± 0.04	0.78 ^{aA} ± 0.13
Adhesiveness, mJ	1	1.10 ^{aA} ± 0.95	1.73 ^{bB} ± 0.19	1.77 ^{bA} ± 0.21	1.70 ^{bB} ± 0.26
	21	1.04 ^{bA} ± 0.17	1.13 ^{bA} ± 0.13	1.83 ^{cA} ± 0.05	0.67 ^{aA} ± 0.06
Cohesiveness	1	0.63 ^{aA} ± 0.19	0.55 ^{aA} ± 0.21	0.46 ^{aA} ± 0.02	0.63 ^{aA} ± 0.22
	21	0.56 ^{aA} ± 0.09	0.50 ^{aA} ± 0.07	0.44 ^{aA} ± 0.12	0.59 ^{aA} ± 0.06
Springiness, mm	1	13.64 ^{aA} ± 0.54	14.74 ^{aA} ± 0.67	13.45 ^{aA} ± 0.27	13.94 ^{aA} ± 0.53
	21	13.41 ^{aA} ± 0.75	14.02 ^{aA} ± 0.56	13.55^{aA} ± 0.35	13.93^{aA} ± 0.14

[1]—control milk; [2]—fermented milk with rosehip; [3]—fermented milk with acerola; [4]—fermented milk with L(+)ascorbic acid; 1—after 1st day of storage; 21—after 21st day of storage; Values are mean ± SD; n = 15 for each group; A,B—mean values in columns denoted by different letters differ statistically significantly ($p < 0.05$) depending on the storage time; a–c—mean values in lines denoted by different letters differ statistically significantly ($p < 0.05$) depending on the different forms of vitamin C.

The DR milk with rosehip had the highest hardness, which may be due to the pectin polysaccharide present in organic rosehip powder. According to Ognyanov et al. [51], this polysaccharide fraction consists mainly of galacturonic acid (45.5%), galactose (5.5%) and arabinose (4.7%). Presumably, homogalacturonan, which is the primary building block of pectin in rosehip consisting of long sequences of a molecule composed of 1,4-linked α-d-galacturonic acid (GalA) interspersed with small blocks of non-methyl-esterified GalA units, causes an increase in the hardness of the tested DR milk [52].

The addition of ascorbic acid to VC milk resulted in a gel with the lowest hardness on the 1st and 21st days of storage.

These results agreed with Sah et al. [53] and Tudorica et al. [54], who observed lower hardness or consistency of yoghurt with selected additives. The lower hardness could be related to structural changes in the acid gel with loss of firmness of the protein matrix [55–57]. The addition of vitamin C from different sources induced a change in the acid gel structure. Furthermore, Sah et al. [53] found that the decrease in firmness values in yoghurt with pineapple fiber powder could be attributed to incompatibility between milk proteins and fruit fiber polysaccharides, which was not confirmed by our study as the addition of wild rose and acerola increased hardness.

Cohesiveness measures the degree of compression to which a sample is compressed between teeth before breaking, which is the resistance of the internal connections [58]. In the present study, cohesiveness values ranged from 0.44 to 0.63. As for springiness, there was no significant effect of added vitamin C from different sources or storage time on these texture components.

2.6. Microbiology Analysis

The viable cell counts of *Lactobacillus rhamnosus* on the 1st and 21st day of storage are shown in Figure 1. The growth of *L. rhamnosus* bacteria was enhanced by adding rosehip to the milk before fermentation. After the 1st day of storage of DR milk, the viable cell count of *L. rhamnosus* bacteria was 0.18 log cfu g^{-1} higher than the control. In the other milk samples with added vitamin C (AC and VC), the number of viable *L. rhamnosus* bacterial cells was lower compared to the control, but the differences were not significant. After the first day of storage, bacterial cell counts above 9 log cfu g^{-1} were determined in all fermented milk samples. The results of Choi and Lim [59] suggest that additional enrichment with 0.025% ascorbic acid and 0.5% yeast extract of yoghurt supplemented with 0.5% EBG (enzyme-bioconverted ginseng) may be beneficial to produce probiotic yoghurt with high amounts of *L. acidophilus* and *Bifidobacterium*. In our study, only the addition of rosehip had a stimulating effect on the growth of *L. rhamnosus*. In contrast, the addition of acerola and ascorbic acid caused a decrease in the number of bacterial cells compared to the control after one day of storage.

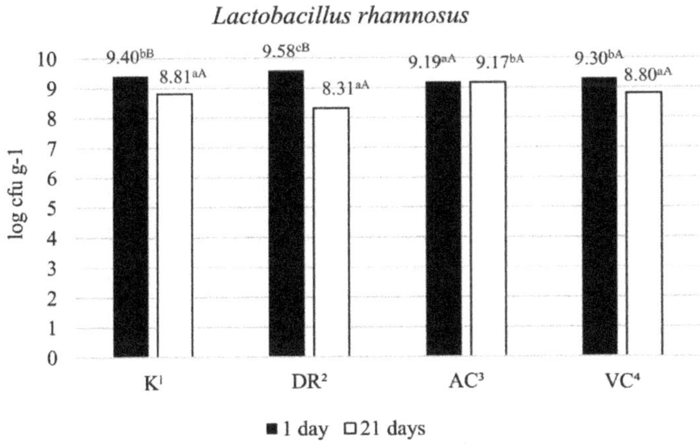

Figure 1. Viable counts in milk fermented by *L. rhamnosus* depending on the source of vitamin C [log cfu g^{-1}]; Values are means ± SD; n = 15 for each group; [1]—control milk; [2]—fermented milk with rosehip; [3]—fermented milk with acerola; [4] —fermented milk with L(+)ascorbic acid; A,B—mean values between 1 and 21 days of storage time denoted by different letters differ statistically significantly ($p < 0.05$); a–c—mean values between the form of vitamin C denoted by different letters differ statistically significantly ($p < 0.05$).

The number of *L. rhamnosus* cells during storage was reduced by extending the storage time to 21 days (Figure 1), especially in DR milk (by 1.2 log cfu g^{-1}), VC milk (by 0.5 log cfu g^{-1}) and control milk by (0.6 log cfu g^{-1}). Only in AC fermented milk, a similar number of *L. rhamnosus* cells was determined after 21 days compared to the first day. This indicates that acerola's addition helped increase bacterial cell viability and maintain a constant level of viable bacterial cells throughout the storage period. The lowest viability of *L. rhamnosus* cells was determined in milk with the addition of rosehip DR. These differences are most probably caused by the presence of different proportions of compounds such as amino acids, phenolic compounds, including anthocyanins and flavonoids, as well as carotenoids and vitamins [45].

The highest pH characterized DR milk at both 1 and 21 days of storage compared to the other milk. According to Collins et al. [60], *L. rhamnosus* requires for growth a high level of vitamins including folic acid, riboflavin, niacin, pantothenic acid and the mineral calcium. The optimum initial pH value for growth is in the range of 6.4 to 4.5. Therefore, *L. rhamnosus* in milk with the addition of rosehip provided good conditions for growth. It could be concluded that the decrease in the number of viable *L. rhamnosus* cells could be related to the highest pH after 21 days of storage compared to the other samples. Most probably, this pH value of DR milk was affected by the compounds present in the rosehip.

Ascorbic acid addition to fermented milk might provide a deoxygenating function and help reduce the oxidation-reduction potential necessary for the viability of probiotic bacteria [61]. In a study by Dave and Shah [38], the addition of ascorbic acid to fermented milk reduced its oxygen content and redox potential during 15 and 20 days of storage, resulting in a slower reduction of Lactobacillus cells.

On the 21st day of storage, the number of *L. rhamnosus* cells in control milk and milk with added vitamin C: DR, AC and VC reached >8 log cfu g^{-1}; thus, these milks met the therapeutic minimum criterion (International Dairy Federation's Recommendation) [62] for probiotic products, which should contain at least 7 log cfu g^{-1} of lactic acid bacteria.

2.7. Organoleptic Evaluation

The results of the organoleptic evaluation are shown in Figures 2 and 3. It was found that the addition of vitamin C from rosehips and acerola improved the consistency of fermented milk and decreased the intensity of sour taste in both evaluation dates compared to control milk. The addition of rosehip and acerola provided the fermented milks with a characteristic additive taste and aroma. According to the assessors, the taste and odor given by the addition of rosehip were the most intense of all the vitamin C sources used in the experiments. Milk with the addition of ascorbic acid VC was characterized by the most intense sour taste and a slightly perceptible smell of fermentation. The K, DR, AC, VC milks did not show any off flavor or off odor in both periods, which was confirmed by the analysis of variance (Table 2).

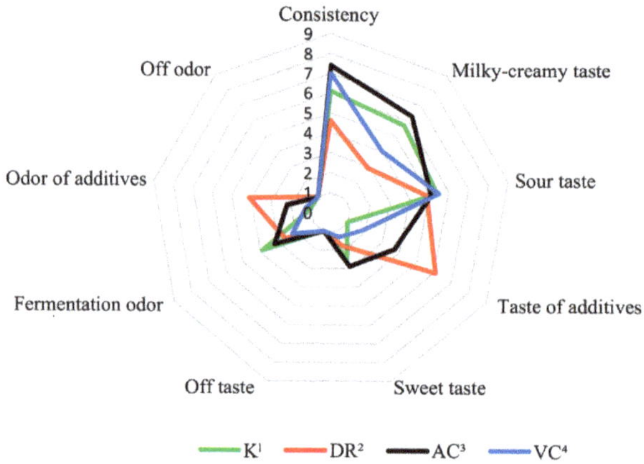

Figure 2. Effect of added different source of vitamin C on organoleptic parameters of fermented milk by *L. rhamnosus* after 1st day of storage; [1]—control milk; [2]—fermented milk with rosehip; [3]—fermented milk with acerola; [4]—fermented milk with L(+)ascorbic acid.

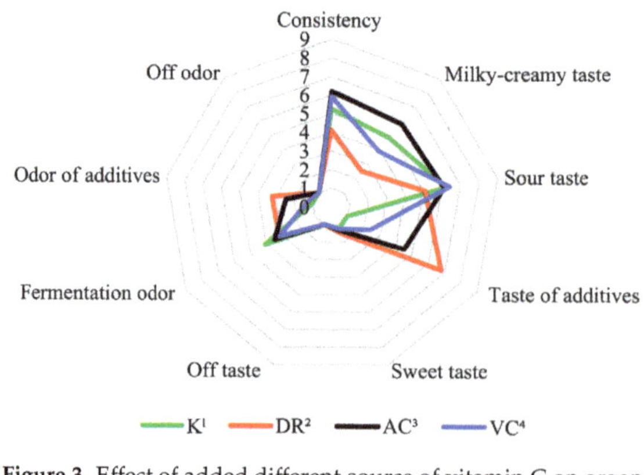

Figure 3. Effect of added different source of vitamin C on organoleptic parameters of fermented milk by *L. rhamnosus* after 21st day of storage; [1]—control milk; [2]—fermented milk with rosehip; [3]—fermented milk with acerola; [4]—fermented milk with L(+)ascorbic acid.

3. Material and Methods

3.1. Materials

Fermented milk was produced from Łaciate milk (2% fat) from SM Mlekpol (Grajewo, Poland). Organic powder of rosehips (Rosa canina L.) (400 mg of vitamin C $\times$ 100 g^{-1}) came from RADZIOWI SP. z.o.o. (Częstochowa, Poland). Acerola in powder form (25,000 mg vitamin C $\times$ 100 g^{-1}) was purchased from Aliness Health'n'beauty (Karczew, Poland) and L(+) ascorbic acid from Chempur (Piekary Śląskie, Poland). The starter culture of probiotic bacteria (*Lactobacillus rhamnosus*) was purchased from Chr. Hansen (Hoersholm, Denmark). MRS agars and peptone water came from Biocorp (Warszawa, Poland). Sodium hydroxide, phenolphthalein, ascorbic acid, sodium bicarbonate, oxalic acid, 2,6-dichlorophenolindophenol, were purchased from Chempur (Piekary Śląskie, Poland).

All of the reagents used were of analytical reagent grade.

Preliminary Studies

The amount of vitamin C addition depending on the source of origin was determined based on the preliminary experiments. The addition of higher doses of vitamin C depending on the source, to milk at 37 °C before fermentation caused denaturation of milk proteins, which also made further studies impossible. The applied amount of vitamin C was the maximum possible dose that did not cause milk protein denaturation.

3.2. Preparation of Fermented Milk

The milk was pasteurized at 85 °C, 30 min, then cooled to 40 °C and divided into groups:

K—control milk;
AC—fermented milk with acerola;
DR—fermented milk with rosehip;
VC—fermented milk with L(+) ascorbic acid.

Next, the milk was inoculated with a single starter culture of Lactobacillus rhamsosus (Chr. Hansen, Hvidovre, Denmark). The control sample was the milk without the addition of vitamin C. Each batch of milk was inoculated with a previously activated starter culture *Lactobacillus rhamnosus* (in the form of bulk activated at 40 °C for 5 h, which was added to the milk in the amount of 5%), according to the Szajnar et al. [63] method. Inoculated milk was stirred and poured into 100 mL plastic cups and fermented at 37 °C for 10 h. The final products were cooled to 5 °C (Cooled Incubator ILW 115, POL-EKO Aparatura, Wodzisław Śląski, Poland). The experiment was repeated three times. Fermented milk was evaluated after the first and twenty-one days of cold storage (5 °C).

3.3. Determination of Acidity

The pH value was performed with a pH-meter (FiveEasy Mettler Toledo, Greifensee, Switzerland) using an electrode InLab®Solids Pro-ISM (Mettler Toledo, Greifensee, Switzerland).

The milk's total acidity (TA) (g of lactic acid L^{-1}) was determined according to the method Jemaa et al. [64].

3.4. Microbiological Analysis

Each batch of milk was inoculated with a previously activated starter culture *Lactobacillus rhamnosus* (in the form of bulk activated at 40 °C for 5 h, which was added to the milk in the amount of 5%), according to the Lima et al. [65] method. Inoculated milk was stirred and poured into 100 mL plastic cups and fermented at 37 °C for 10 h.

3.5. Syneresis

Syneresis was measured as the amount of whey released relative to the initial weight and averaged five determinations. The 10 g of fermented milk sample was transferred into 50 mL plastic tube and centrifuged Refrigerated Centrifuge LMC-4200R (Biosan SIA, Riga, Latvia) [66].

3.6. Color of Fermented Milk

The color of fermented milk was analyzed with a colorimeter (the Precision Colorimeter, Model NR 145, Shenzhen, China) using the instrumental method and the CIELAB system. The image lightness was determined with the parameter L* and chromaticity using a*, b*, C, h^0. Prior to the measurement, the device was calibrated on a white reference standard [67].

3.7. Determination of Vitamin C Content

The content of ascorbic acid was determined in the samples by Tillimans' titration method [PN-A-04019:1998] [68], which is based on the extraction of ascorbic acid from the product with oxalic acid, followed by its oxidation to dehydroascorbic acid in an acidic medium using the titrated blue dye 2,6-dichlorophenolindophenol (DCIP).

3.8. Parameters of Texture

Texture profile was determined with TPA test using CT3 Texture Analyzer (Brookfield, Middleboro, MA, USA) with Texture Pro CT (Brookfield, Middleboro, MA, USA) software. The dimensions of the sample were: 66 mm × 33.86 mm (cylinder), and the sample temperature was 8 °C. The experiment was conducted using the acrylic probe TA 3/100 with the following settings: distance 15 mm, contact load 0.1 N, measurement speed 1 mm/s [67].

3.9. Organoleptic Evaluation

The analysis was made for three fermented milk samples with vitamin C and the control sample. The organoleptic evaluation was performed by a trained panel (15 women and 15 men, the age of 25–30). The panelists were served four samples at a time and asked to rinse their mouths with water between samples. The samples of fermented milk were assessed on a 9-point rating scale with edge markings. The left end denoted the least intense, the least characteristic feature. The panelists evaluated the presence of sandy and grainy consistency; impalpable milky-creamy taste, impalpable sour taste, impalpable fermentation odor, impalpable sweet taste, impalpable the off taste and odor, impalpable additives taste and odor, and the right end denoted the most characteristic feature: smooth texture; the most intense milky-creamy taste, the most intense sour taste, the most intense fermentation odor, the most intense sweet taste, the most intense the off taste and odor, the most intense characteristic taste and odor of additives [69], PN-ISO 22935-2:2013-07 [70].

Definition of the attributes in the descriptive organoleptic analysis of fermented milk:

Milky-creamy taste: the taste stimulated by milk powder.
Sour taste: the taste stimulated by lactic acid.
Taste of additives: the taste stimulated by added vitamin C depending on the source of origin.
Sweet taste: the taste stimulated by sucrose.
Off-taste: an unidentified taste that is not characteristic.
Fermentation odor: the intensity of odor associated with sour milk, i.e., lactic acid.
Odor of additives: odor characteristic stimulated by added vitamin C depending on the source of origin.
Off-odor: unidentified odor that is not characteristic.

3.10. Statistical Analysis

The results from three independent studies were expressed as the mean and standard deviation in Statistica v. 13.1 (StatSoft, Tulsa, OK, USA). One and two-way ANOVA was performed, and the differences between the mean values were verified with the Turkey test, with $p < 0.05$.

4. Conclusions

This study showed good stability of vitamin C from different sources (rosehip, acerola and ascorbic acid in powder form) in fermented milk, as losses were only 5–8%. Depending on the vitamin C source, the color, hardness and organoleptic characteristics of the probiotic milk were significantly affected. After 21 days of storage, all tested milks contained >8 log cfu g^{-1} *L. rhamnosus*, which provided a therapeutic effect. Due to the advantages of natural products and favorable consumer perception, the addition of rosehip and acerola could be considered promising candidates for developing the functional food market.

Author Contributions: Conceptualization, K.S. and A.Z.; methodology, K.S. and A.Z.; software, K.S., A.Z.; validation, K.S., A.Z. and M.P.; formal analysis, K.S., A.Z.; investigation, K.S., A.Z. and M.P..; data curation, A.Z.; writing—original draft preparation, K.S., A.Z. and M.P.; writing—review and editing, K.S. and M.P.; visualization, K.S.; supervision, A.Z.; project administration, K.S. funding acquisition, K.S. All authors have read and agreed to the published version of the manuscript.

Funding: The project is financed by the program of the Minister of Science and Higher Education named "Regional Initiative of Excellence" in the years 2019–2022, project number 026/RID/2018/19, the amount of financing PLN 9 542 500.00.

Institutional Review Board Statement: Not applicable.

Informed Consent Statement: Not applicable.

Data Availability Statement: Not applicable.

Conflicts of Interest: The authors declare no conflict of interest.

Sample Availability: Samples of the compounds are not available from the authors.

References

1. Cremonini, F.; Di Caro, S.; Nista, E.C.; Bartolozzi, F.; Capelli, G.; Gasbarrini, G.; Gasbarrini, A. Meta-analysis: The effect of probiotic administration on antibiotic-associated diarrhea. *Aliment. Pharmacol. Ther.* **2002**, *16*, 1461–1467. [CrossRef]
2. Johnston, B.C.; Supina, A.L.; Ospina, M.; Vohra, S. Probiotics for the prevention of pediatric antibotic-associated diarrhea. *Cochrane Database Syst. Rev.* **2007**, *18*, CD004827. [CrossRef]
3. Lee, J.; Seto, D.; Bielory, L. Meta-analysis of clinical trials of probiotics for prevention and treatment of pedatric atopic dermatitis. *J. Allergy Clin. Immunol.* **2008**, *121*, 116–121. [CrossRef]
4. Oleksy-Sobczak, M.; Klewicka, E.; Piekarska-Radzik, L. Exopolysaccharides production by *Lactobacillus rhamnosus* strains–Optimization of synthesis and extraction conditions. *LWT-Food Sci. Technol.* **2020**, *122*, 109055. [CrossRef]
5. Sharareh, H.; Soltani, H.; Reid, G. Growth and survival of *Lactobacillus* reuteri RC-14 and *Lactobacillus rhamnosus* GR-1 in yogurt for use as a functional food. *Innov. Food Sci. Emerg. Technol.* **2009**, *10*, 293–296. [CrossRef]
6. Jyoti, B.D.; Suresh, A.K.; Venkatesh, K.V. Effect of preculturing conditions on growth of *Lactobacillus rhamnosus*, on medium containing glucose and citrate. *Microbiol. Res.* **2004**, *159*, 35–42. [CrossRef] [PubMed]
7. Innocente, N.; Biasutti, M.; Rita, F.; Brichese, R.; Comi, G.; Iacumin, L. Effect of indigenous *Lactobacillus rhamnosus* isolated from bovine milk on microbiological characteristics and aromatic profile of traditional yogurt, *Lebensm. Wiss. Technol.* **2016**, *66*, 158–164. [CrossRef]
8. Kamal, R.M.; Alnakip, M.E.; Abd El Aal, S.F.; Bayoumi, M.A. Bio-controlling capability of probiotic strain *Lactobacillus rhamnosus* against some common foodborne pathogens in yoghurt. *Int. Dairy J.* **2018**, *85*, 1–7. [CrossRef]
9. Maćkowiak, K.; Torliński, L. Współczesne poglądy na rolę witaminy C w filozjologii i patologii człowieka. Contemporary view on the role of vitamin c in human physiology and pathology. *Now. Lek.* **2007**, *76*, 349–356. (In Polish)
10. Miktus, M. Witaminy część II: Ogólna charakterystyka witaminy C. Vitamins part II: General characteristics of vitamin C. *Zyw. i Zdrowie.* **2000**, *3*, 1–4. (In Polish)
11. Kleszczewska, E. Biologiczne znaczenie witaminy C ze szczególnym uwzględnieniem jej znaczenia w metabolizmie skóry. Biological role of vitamin C and importance in the skin metabolism. *Pol. Merkur. Lek.* **2007**, *138*, 462–465. (In Polish)
12. Zhang, P.Y.; Xu, X.; Li, X.C. Cardiovascular diseases: Oxidative damage and antioxidant protection. *Eur. Rev. Med. Pharmacol. Sci.* **2014**, *18*, 3091–3096.
13. Prakash, A.; Baskaran, R. Acerola, an untapped functional superfruit: A review on latest frontiers. *J. Food Sci. Technol.* **2018**, *55*, 3373–3384. [CrossRef]

14. Marques, T.R.; Caetano, A.A.; Simão, A.A.; Castro, F.C.D.O.; Ramos, V.D.O.; Corrêa, A.D. Methanolic extract of Malpighia emarginata bagasse: Phenolic compounds and inhibitory potential on digestive enzymes. *Rev. Bras. Farmacogn.* **2016**, *26*, 191–196. [CrossRef]
15. Hanamura, T.; Uchida, E.; Aoki, H. Changes of the composition in acerola (Malpighia emarginata DC.) fruit in relation to cultivar, growing region and maturity. *J. Sci. Food Agric.* **2008**, *88*, 1813–1820. [CrossRef]
16. Hanamura, T.; Uchida, E.; Aoki, H. Skin-lightening effect of a polyphenol extract from Acerola (Malpighia emarginata DC.) fruit on UV-induced pigmentation. *Biosci. Biotechnol. Biochem.* **2008**, *72*, 3211–3218. [CrossRef] [PubMed]
17. Righetto, A.M.; Netto, F.M.; Carraro, F. Chemical composition and antioxidant activity of juices from mature and immature acerola (Malpighia emarginata DC). *Food Sci. Technol. Int.* **2005**, *11*, 315–321. [CrossRef]
18. Vendramini, A.L.; Trugo, L.C. Chemical composition of acerola fruit (Malpighia punicifolia L.) at three stages of maturity. *Food Chem.* **2000**, *71*, 195–198. [CrossRef]
19. Mármol, I.; Sánchez-de-Diego, C.; Jiménez-Moreno, N.; Ancín-Azpilicueta, C.; Rodríguez-Yoldi, M.J. Therapeutic Applications of Rose Hips from Different Rosa Species. *Int. J. Mol. Sci.* **2017**, *18*, 1137. [CrossRef]
20. Hvattum, E. Determination of phenolic compounds in rose hip (Rosa canina) using liquid chromatography soupled to electrospray ionization tandem mass spectrometry and diode-array detection. *Rapid Commun. Mass Spetrometry* **2002**, *16*, 655–662. [CrossRef]
21. Adamczak, A.; Buchwald, W.; Zieliński, J.; Mielcarek, S. Flavonoid and organic acid content in rose hips (Rosa, L., sect. Caninae DC. EM. Christ.). *Acta Biol. Crac. Ser. Bot.* **2012**, *54*, 105–112. [CrossRef]
22. Nojavan, S.; Khalilian, F.; Kiaie, F.M.; Rahimi, A.; Arabanian, A.; Chalavi, S. Extraction and quantitative determination of ascorbic acid during different maturity stages of Rosa canina L. fruit. *J. Food Compos. Anal.* **2008**, *21*, 300–305. [CrossRef]
23. Ball, G.F.M. *Vitamins in Foods/Analysis, Bioavailability, and Stability*; CRC Press Taylor & Francis Group: Boca Raton, FL, USA, 2006
24. Linares, D.; Michaud, P.; Delort, A.M.; Traïkia, M.; Warrand, J. Catabolism of L-ascorbate by *Lactobacillus rhamnosus* GG. *J. Agric. Food Chem.* **2011**, *59*, 4140–4147. [CrossRef]
25. Pérez-Vicente, A.; Gil-Izquierdo, A.; García-Viguera, C. In vitro gastrointestinal digestion study of pomegranate juice phenolic compounds, anthocyanins, and vitamin C. *J. Agric. Food Chem.* **2002**, *50*, 2308–2312. [CrossRef]
26. Yuan, J.P.; Chen, F. Degradation of ascorbic acid in aqueous solution. *J. Agric. Food Chem.* **1998**, *46*, 5078–5082. [CrossRef]
27. Budsławski, J. *Zarys chemii mleka. Outline of milk chemistry Państwowe Wydawnictwo Rolnicze i Leśne*; PWRiL: Warszawa, Poland, 1971. (In Polish)
28. Klupsch, H.J.; Verlag, T. *Acid Milk Products, Milk Beverage and Desserts*; Verlag Th. Mann: Gelsenkirchen–Buer, Germany, 1992.
29. Ogah, O.; Watkins, C.S.; Ubi, B.E.; Oraguzie, N.C. Phenolic compounds in Rosaceae fruit and nut crops. *J. Agric. Food Chem.* **2014**, *62*, 9369–9386. [CrossRef] [PubMed]
30. Demir, N.; Yildiz, O.; Alpaslan, M.; Hayaloglu, A. Evaluation of volatiles, phenolic compounds and antioxidant activities of rose hip (Rosa L.) fruits in Turkey. *LWT-Food Sci. Technol.* **2014**, *57*, 126–133. [CrossRef]
31. Elmastas, M.; Demir, A.; Genc, N.; Dölek, Ü.; Günes, M. Changes in flavonoid and phenolic acid contents in some Rosa species during ripening. *Food Chem.* **2017**, *235*, 154–159. [CrossRef] [PubMed]
32. Nadpal, J.D.; Lesjak, M.M.; Šibul, F.S.; Anackov, G.T.; Cetojevic'-Simin, D.D.; Mimica-Dukic, N.M.; Beara, I.N. Comparative study of biological activities and phytochemical composition of two rose hips and their preserves: Rosa canina L. and Rosa arvensis Huds. *Food Chem.* **2016**, *192*, 907–914. [CrossRef]
33. Tumbas, V.T.; Canadanovic-Brunet, J.M.; Cetojevic-Simin, D.D.; Cetkovic, G.S.; Ethilas, S.M.; Gille, L. Effect of rosehip (Rosa canina L.) phytochemicals on stable free radicals and human cancer cells. *J. Sci. Food Agric.* **2012**, *92*, 1273–1281. [CrossRef]
34. Hosni, K.; Chrif, R.; Zahed, N.; Abid, I.; Medfei, W.; Sebei, H.; Brahim, N.B. Fatty acid and phenolic constituents of leaves, flowers and fruits of tunisian dog rose (Rosa canina L.). *Riv. Ital. Sostanze Grasse* **2010**, *87*, 117–123.
35. Türkben, C.; Uylaşer, V.; İncedayı, B.; Çelikkol, I. Effects of different maturity periods and processes on nutritional components of rose hip (Rosa canina L.). *J. Food Agric. Environ.* **2010**, *8*, 26–30.
36. Olsson, M.E.; Gustavsson, K.E.; Andersson, S.; Nilsson, A.; Duan, R.D. Inhibition of cancer cell proliferation in vitro by fruit and berry extracts and correlations with antioxidant levels. *J. Agric. Food Chem.* **2004**, *52*, 7264–7271. [CrossRef] [PubMed]
37. Ilic, D.B.; Ashoor, S.H. Stability of Vitamins A and C in Fortified Yogurt. *J. Dairy Sci.* **1988**, *71*, 1492–1498. [CrossRef]
38. Dave, R.I.; Shah, N.P. Effectiveness of ascorbic acid as an oxygen scavenger in improving viability of probiotic bacteria in yoghurts made with commercial starter cultures. *Int. Dairy J.* **1997**, *7*, 435–443. [CrossRef]
39. Graulet, B. Ruminant milk: A source of vitamins in human nutrition. *Anim. Front.* **2014**, *4*, 24–30. [CrossRef]
40. Rani, R.; Dharaiya, C.N.; Unnikrishnan, V.; Singh, B. Factors Affecting Syneresis from Yoghurt for Preparation of Chakka. *Indian J. Dairy Sci.* **2012**, *65*, 135–140.
41. Peng, Y.; Horne, D.S.; Lucey, J.A. Impact of preacidification of milk and fermentation time on the properties of yogurt. *J. Dairy Sci.* **2009**, *92*, 2977–2990. [CrossRef]
42. Oktavia, H.; Radiati, L.E.; Rosyidi, D. Evaluation of Physicochemical Properties and Exopolysaccharides Production of Single Culture and Mixed Culture in Set Yoghurt. *J-PAL* **2016**, *7*, 52–59.
43. Sun, J.; Chen, H.; Qiao, Y.; Liu, G.; Leng, C.; Zhang, Y.; Lv, X.; Feng, Z. The nutrient requirements of *Lactobacillus rhamnosus* GG and their application to fermented milk. *J. Dairy Sci.* **2019**, *102*, 5971–5978. [CrossRef]

44. Kaszuba, M.; Viapiana, A.; Wesołowski, M. Dzika róża (Rosa canina L.) jako źródło witamin i przeciwutleniaczy w diecie człowieka. Rose hip (Rosa canina L.) as a vitamin and antioxidants source in human diet. *Pol. Tow. Farm.* **2019**, *75*, 97–102. (In Polish) [CrossRef]
45. Belwal, T.; Devkota, H.P.; Hassan, H.A.; Ahluwalia, S.; Ramadan, M.F.; Mocan, A.; Atanasov, A.G. Phytopharmacology of Acerola (Malpighia spp.) and its potential as functional food. *Trends Food Sci. Technol.* **2018**, *74*, 99–106. [CrossRef]
46. Lin, D.; Xiao, M.; Zhao, J.; Li, Z.; Xing, B.; Li, X.; Kong, M.; Li, L.; Zhang, Q.; Liu, Y.; et al. An Overview of Plant Phenolic Compounds and Their Importance in Human Nutrition and Management of Type 2 Diabetes. *Molecules* **2016**, *21*, 1374. [CrossRef]
47. Damin, M.R.; Alcântara, M.R.; Nunes, A.P.; Oliveira, M.N. Effects of milk supplementation with skim milk powder, whey protein concentrate and sodium caseinate on acidification kinetics, rheological properties and structure of nonfat stirred yogurt. *LWT-Food Sci. Technol.* **2009**, *42*, 1744–1750. [CrossRef]
48. Macit, E.; Bakirci, İ. Effect of different stablizers on quality characteristics of the set-type yoğurt. *Afr. J. Biotechnol.* **2017**, *16*, 2142–2151. [CrossRef]
49. Bourne, M. *Food Texture and Viscosity Concept and Measurement*; Academic Press: New York, NY, USA, 2002.
50. Mousavi, M.; Heshmati, A.; Daraei Garmakhany, A.; Vahidinia, A.; Taheri, M. Texture and sensory characterization of functional yogurt supplemented with flaxseed during cold storage. *Food Sci. Nutr.* **2019**, *7*, 907–917. [CrossRef]
51. Ognyanova, M.; Remoroza, C.; Schols, H.A.; Georgiev, Y.; Kratchanova, M.; Kratchanov, C. Isolation and structure elucidation of pectic polysaccharide from rose hip fruits (Rosa canina L.). *Carbohydr. Polym.* **2016**, *151*, 803–811. [CrossRef]
52. Albersheim, P.; Darvill, A.G.; O'Neill, M.A.; Schols, H.A.; Voragen, A.G.J. *An Hypothesis: The Same Six Polysaccharides are Components of the Primary Cell Walls of All Higher Plants*; Visser, J., Voragen, A.G.J., Eds.; Elsevier: Amsterdam, The Netherlands, 1996; pp. 47–55.
53. Sah, B.N.P.; Vasiljevic, T.; McKechnie, S.; Donkor, O.N. Physicochemical, textural and rheological properties of probiotic yogurt fortified with fibre-rich pineapple peel powder during refrigerated storage. *LWT-Food Sci. Technol.* **2016**, *65*, 978–986. [CrossRef]
54. Tudorica, C.M.; Jones, T.E.R.; Kuri, V.; Brennan, C.S. The effects of refined barley beta-glucan on the physico-structural properties of low-fat dairy products: Curd yield, microstructure, texture and rheology. *J. Sci. Food Agric.* **2004**, *84*, 1159–1169. [CrossRef]
55. Lee, W.J.; Lucey, J.A. Formation and physical properties of yogurt. *Asian-Australas. J. Anim. Sci.* **2010**, *23*, 1127–1136. [CrossRef]
56. Santillan-Urquiza, E.; Mendez-Rojas, M.A.; Velez-Ruiz, J.F. Fortification of yogurt with nano and micro sized calcium, iron and zinc, effect on the physicochemical and rheological properties. *LWT-Food Sci. Technol.* **2017**, *80*, 462–469. [CrossRef]
57. Tan, P.Y.; Tan, T.B.; Chang, H.W.; Tey, B.T.; Chan, E.S.; Lai, O.M.; Baharin, B.S.; Nehdi, I.A.; Tan, C.P. Effects of storage and yogurt matrix on the stability of tocotrienols encapsulated in chitosan-alginate microcapsules. *Food Chem.* **2018**, *241*, 79–85. [CrossRef]
58. Curti, C.A.; Vidal, P.M.; Curti, R.N.; Ramon, A.N. Chemical characterization, texture and consumer acceptability of yogurts supplemented with quinoa flour. *Food Sci. Technol.* **2017**, *37*, 627–631. [CrossRef]
59. Choi, S.H.; Lim, Y.S. Viability of Probiotic Bacteria in Yogurt Supplemented with Enzyme-Bioconverted Ginseng, Ascorbic Acid, and Yeast Extract. *J. Dairy Sci. Biotechnol.* **2019**, *37*, 57–68. [CrossRef]
60. Collins, F.L.; Rios-Arce, N.D.; Schepper, J.D.; Parameswaran, N.; McCabe, L.R. The Potential of Probiotics as a Therapy for Osteoporosis. *Microbiol. Spectr.* **2017**, *5*, 4. [CrossRef] [PubMed]
61. Talwalkar, A.; Kailasapathy, K. A review of oxygen toxicity in probiotic yogurts: Influence on the survival of probiotic bacteria and protective techniques. *Compr. Rev. Food Sci. Food Saf.* **2004**, *3*, 117–124. [CrossRef]
62. Champagne, C.P.; . Gardner, N.J.; Roy, D. Challenges in the Addition of Probiotic Cultures to Foods. *Crit. Rev. Food Sci. Nutr.* **2005**, *45*, 61–84. [CrossRef]
63. Szajnar, K.; Znamirowska, A.; Kuźniar, P. Sensory and textural properties of fermented milk with viability of *Lactobacillus rhamnosus* and *Bifidobacterium animalis* ssp. *lactis* Bb-12 and increased calcium concentration. *Int. J. Food Prop.* **2020**, *23*, 582–598. [CrossRef]
64. Jemaa, M.B.; Falleh, H.; Neves, M.A.; Isoda, H.; Nakajima, M.; Ksouri, R. Quality Preservation of Deliberately Contaminated Milk Using Thyme Free and Nanoemulsified Essential Oils. *Food Chem.* **2017**, *217*, 726–734. [CrossRef]
65. Lima, K.G.; Kruger, M.F.; Behrens, J.; Destro, M.T.; Landgraf, M.; Franco, B.D.G. Evaluation of Culture Media for Enumeration of *Lactobacillus acidophilus, Lactobacillus casei* and *Bifidobacterium animalis* in the presence of *Lactobacillus delbrueckii* subsp *bulgaricus* and *Streptococcus thermophilus*. *LWT-Food Sci. Technol.* **2009**, *42*, 491–495. [CrossRef]
66. Szajnar, K.; Znamirowska, A.; Kalicka, D. Effects of various magnesium salts for the production of milk fermented by Bifidobacterium animalis ssp. lactis Bb-12. *Int. J. Food Prop.* **2019**, *22*, 1087–1099. [CrossRef]
67. Szajnar, K.; Pawlos, M.; Znamirowska, A. The Effect of the Addition of Chokeberry Fiber on the Quality of Sheep's Milk Fermented by *Lactobacillus rhamnosus* and *Lactobacillus acidophilus*. *Int. J. Food Sci.* **2021**, *1*, 1–9. [CrossRef] [PubMed]
68. PN-A-04019:1998. In *Produkty Spożywcze—Oznaczanie Zawartości Witaminy C. Food Products—Determination of Vitamin C*; Polish Committee for Standardization: Warsaw, Poland, 1998. (In Polish)
69. Baryłko-Pikielna, N.; Matuszewska, I. Sensoryczne Badania Żywności. Podstawy—Metody—Zastosowania. Sensory Food Testing. Fundamentals-Methods-Applications. *Wyd. Nauk. Pttż. Krakow.* **2014**, *66*, 150–157. (In Polish)
70. PN-ISO 22935-2:2013-07. In *Milk and Milk Products—Sensory Analysis—Part. 2: Recommended Methods for Sensory Evaluation*; Polish Committee for Standardization: Warsaw, Poland, 2013. (In Polish)

molecules

MDPI

Article

Changes in the Folate Content and Fatty Acid Profile in Fermented Milk Produced with Different Starter Cultures during Storage

Marta Czarnowska-Kujawska * and Beata Paszczyk

Department of Commodity and Food Analysis, The Faculty of Food Sciences, University of Warmia and Mazury in Olsztyn, 10-726 Olsztyn, Poland; paszczyk@uwm.edu.pl
* Correspondence: marta.czarnowska@uwm.edu.pl; Tel.: +48-89-524-52-76

Abstract: The application of bacterial cultures in food fermentation is a novel strategy to increase the "natural" levels of bioactive compounds. The unique ability of lactic acid bacteria (LAB) to produce folate, B vitamins, and conjugated linolenic acid *cis9trans*11 C18:2 (CLA) during cold storage up to 21 days was studied. Although some species of LAB can produce folates and other important nutrients, little is known about the production ability of yogurt starter cultures. Pasteurized milk samples were inoculated with four different combinations of commercially available yogurt vaccines, including starter cultures of *Bifidobacterium bifidum*. Both the type of vaccine and the time of storage at 8 °C had a significant effect on the folate and CLA contents in the tested fermented milks. The highest folate content (105.4 μg/kg) was found in fresh fermented milk inoculated with *Lactobacillus delbrueckii*, *Streptococcus thermophilus*, and *Bifidobacterium bifidum*. Only the mix of *Lactobacillus delbrueckii* subsp. *bulgaricus*, *Streptococcus thermophilus*, and *Bifidobacterium bifidum* showed potential (59% increase) to synthesize folate during seven days of storage. A significant increase in the content of CLA, when compared to fresh fermented milk, was observed during cold storage for up to 21 days in products enriched with *Bifidobacterium bifidum*.

Keywords: dairy products; biofortification; probiotic; LAB; folic acid; fatty acids; CLA; storage time

1. Introduction

Improving the nutritional value of food is one of the major challenges for the food industry of the twenty-first century. Consumers, apart from new tastes and flavors, are increasingly interested in the origin and bioactive properties of foods that can affect their health and wellbeing [1,2]. Folates are reduced folic acid derivatives (so-called polyglutamates) which naturally occur in food, both of plant and animal origin. The name folic acid refers only to the synthetic form of the vitamin, which is commonly used for food fortification and supplementation [1,3]. Folates cannot be synthesized by humans and must be obtained exogenously. The richest sources are animal liver, yeast, wheat germ, and green leafy vegetables, such as spinach, broccoli and asparagus, and pulses [4]. Folates belong to the water-soluble B vitamins group and are essential components of the human diet for the synthesis reaction of nucleotides and vitamins. They are the source of methyl groups in the process of homocysteine remethylation to methionine [1,5]. Despite such an important role, acquired folate deficiency is common and affects billions of people worldwide, both in developing and developed countries and in different age groups. This micronutrient deficiency is associated with poor diet, malabsorption, alcohol consumption, obesity, and kidney failure [1,4]. An increased risk of folate deficiency especially concerns the elderly (because of lower food intake), children (who are likely to consume a limited variety of food), and pregnant women (because of the critical role of folates in fetus neural tube development). Low folate intake increases the risk of birth defects (spina bifida, anencephaly, encephalocele) [4]. Prolonged insufficient folate consumption is also linked to macrocytic anemia, cardiovascular diseases, neurodegenerative diseases (Alzheimer's, Parkinson's), certain forms of cancers (colorectal, breast, cervical, lung, pancreatic cancer), and even an increased risk of depressive mental disorders [6,7].

In the face of insufficient folate consumption, many developed countries, including the U.S. and Canada, have introduced mandatory folic acid fortification of wheat and/or maize flour at the industrial level to increase the folate intake of the whole population [8,9]. For pregnant women, daily folic acid supplementation is recommended to reduce the risk of low birth weight and congenital malformations, including neural tube defects [10]. However, food fortification programs are not always effective. It is estimated currently that no more than 25% of folic-acid-preventable neural tube defects are actually prevented [9]. Moreover, higher levels of synthetic folic acid in the diet can cause adverse effects, such as masking a vitamin B12 deficiency, which may result in the progression of neuropathy to an irreversible point [11]. Moreover, high physiological folate concentrations and folate overload, as a result of excessive supplementation with folic acid, may increase the risk of impaired brain development in embryogenesis and even possess a growth advantage for pre-cancerous altered cells [4,5]. Therefore, due to the potential risk of using synthetic folic acid, but also limited availability of folate-rich foods (depending on the season and geographic and economic factors), food fortification by natural folate is a necessary alternative [9].

One novel strategy to increase natural folate levels is the application of bacterial cultures in food fermentation. Fermented milk, especially yogurt, is the appropriate choice and ideal matrix for bio-fortification for several reasons. Firstly, it contains folate-binding protein, which improves folates stability and bioavailability. Secondly, it is consumed in many countries worldwide and is increasingly popular among consumers, who consider it healthy, with many therapeutic and desirable effects. Finally, fermented dairy products can significantly contribute to the daily recommended levels by in situ fortifications through fermentation with the aid of folate-producing microbes [9,12,13].

Lactic acid bacteria (LAB) are microorganisms involved in the traditional fermentation process. Some of them are considered probiotic species, which are beneficial to their hosts in many ways and are also known for folate production in high amounts, which may improve the nutritional composition of fermented products [5,14]. The ability of *Streptococcus thermophilus* and *Lactococcus lactis* to synthesize folate has been reported, and increased folate production by *Lactobacillus plantarum* has also been demonstrated. Unfortunately, in the milk fermentation process, the majority of the bacteria are folate utilizers and decrease folate amounts. Therefore, only the proper selection of suitable starter culture or a consortium of folate producers would allow increasing the dietary folate content [12,14–17]. For this reason, further research and development of folate-producing bacteria for food applications should be encouraged.

Besides the ability to produce folates, probiotic bacteria also show the ability to synthesize conjugated linolenic acid *cis*9*trans*11 C18:2 (CLA). This acid is the main representative of the conjugated dienoic milk fat group and constitutes from 75% to over 90% of the sum of these isomers in the fat of milk and dairy products [18–20]. As reported by various authors [21–25], CLA displays a number of health-positive properties, e.g., anticarcinogenic, anti-atherosclerotic, antioxidative, and anti-inflammatory effects. The content of CLA in milk fat may vary widely depending on many factors, such as animal feeding, breed, age, and lactation period [26–32]. In dairy products, it may also be affected by the production process. According to some studies [33–38], technological treatments applied in the industry and additives used may influence CLA content in the fatty acid composition of dairy products. According to literature data [39–44], selected strains of bacteria are capable of CLA synthesis during fermentation. As reported by Kim and Liu [40], CLA content in fermented milk is affected by the type of bacterial strain applied, cell count, appropriate concentration of the substrate, and incubation conditions (time and pH).

In the food additives market, special mixtures of starter cultures that provide good conditions for the right milk fermentation are commercially available. These are mixtures of one or more strains with acid-forming, protective, and organoleptic properties. Producers are expanding their offer with probiotic strains with other properties that preserve or add flavor and are also popular among consumers. The objectives of this study were to analyze the folate content in fermented milk samples produced with the use of commercially available yogurt vaccines, including starter cultures of *Bifidobacterium bifidum*, and to assess the stability of the produced folate during refrigerated storage. Moreover, in the tested fermented milk, the influence of the used starter cultures and the time of cold storage on the fatty acid profile (especially on the content of conjugated linoleic acid *cis*9*trans*11 C18:2 (CLA)) were evaluated.

2. Results and Discussion

2.1. Folates

The folate content in milk after pasteurization, which is a basic material for the production of fermented milk beverages, is not very high when compared to other folate-rich foods. Based on the data obtained using three analytical methods (HPLC, radio protein binding, and microbiological), the folate content determined in cow's milk ranged from 50 to 100 µg/L [5,45,46]. In the authors' previous study [47], the folate content in the pasteurized cow's milk was reported to be 36.9 µg/kg. Raw milk processing, such as pasteurization or UHT, as well as transport conditions and storage time, are known to reduce folate levels. Folates are labile compounds, and increased temperature, exposure to oxidizing agents, UV light and sunlight, unfavorable pH, and interaction with other food ingredients (such as metal cations) enhance folate degradation and/or interconversions [16,48]. Dairy products are often fermented by LAB alone or in combination with other microorganisms before consumption [15]. Several studies have reported an increase in folate content in different foods after fermentation by LAB selected for their biosynthesis capacities [45,46,49–53].

In the current study of fermented milk (FM) samples, two folate forms were identified: 5-methyltetrahydrofolate (5-CH_3-H_4folate) and tetrahydrofolate (H_4folate). H_4folate, 5-CH_3-H_4folate, and the total folate content of fresh and stored tested fermented products are shown in Table 1. The methyl form was dominant, which is in agreement with previous studies [5,16,47,49,54]. The lower content of H_4folate may result from the fact that this vitamer, among other folate vitamers, is likely to oxidize into other folate forms at low pH [9]. *Lactobacillus delbrueckii* subsp. *bulgaricus* and *Streptococcus thermophilus* only accumulated comparable amounts of H_4folate and 5-CH_3-H_4folate in the FM 1, both in fresh material and material stored up to 21 days. In the fresh fermented milk, the H_4folate content ranged from 12.2 in FM 2 inoculated with *Lactococcus lactis and Leuconostoc* to 27.5 µg/kg in FM 3 inoculated with the combination of *Lactobacillus delbrueckii, Streptococcus thermophilus*, and *Bifidobacterium bifidum*. The highest 5-CH_3-H_4folate (84.0 µg/kg) was also found in FM 3, but the lowest (24.4 µg/kg) was in FM 1 inoculated with *Lactobacillus delbrueckii* and *Streptococcus thermophilus*.

Similarly, the lowest total folate content (45.3 µg/kg) was observed in FM 1, while the highest (105.4 µg/kg) was in FM 3. With the exception of product 1, in the analyzed samples, content levels were significantly higher than reported previously for pasteurized milk [47]. An increase of 1.6–2.8 fold was observed in fresh fermented milks 2, 3, and 4.

In the current experiment, during refrigerated storage for at least seven days, only the combination of *Lactobacillus delbrueckii* subsp. *bulgaricus*, *Streptococcus thermophilus*, and *Bifidobacterium bifidum* in FM 4 showed the potential to produce one folate vitamer (5-CH_3-H_4folate). The increase in the methyl form content was 59%, which resulted in a total folate content of 79.2 µg/kg on the seventh day of storage at 8 °C. At the same time, a significant decrease in the H_4folate content (31%) was found. The determined folate losses in FM 4 on days 14 and 21 of refrigerated storage did not exceed 15% and 48%, respectively, and were the lowest compared to other fermented milks at the same storage time. During the refrigerated storage of FM 1–3, a significant decrease ($P < 0.05$) in the folate content was observed on the seventh day of cold storage. The lowest losses of 27% after day 7 were observed in the product FM 3. On day 14 in FMs 1–3, folate losses were 50%, and on day 21 exceeded 65% in both FM 2 and FM 3. When comparing the stability of the two identified folate forms in the tested fermented milk samples during storage at 8 °C, only in FM 2 was clearly higher stability of H_4folate noted. The losses during storage ranged from 23% to 25% compared to 59–75% losses of the methyl form. The observed folate losses in the current experiment during storage may result from the acid pH of fermented foods, in which most folate vitamers are less stable and because folate was consumed by bacteria [9,12].

Table 1. Folate content and losses in fresh and refrigerated fermented milk.

Fermented Milk (FM)	Days of Storage at 8 °C	H4folate (μg/kg)	5-CH3-H4folate(μg/kg)	Total Folates (Sum as Folic Acid) (μg/kg)	Folates Losses during Storage (%)
FM 1 *Lactobacillus delbrueckii* subsp. *bulgaricus*, and *Streptococcus thermophilus*	0	22.1 [1] ± 1.1	24.4 ± 1.3	45.3 [a2] ± 0.3	-
	7	12.3 ± 0.8	15.2 ± 0.8	27.1 [b] ± 0.8	40
	14	8.4 ± 0.2	14.2 ± 0.5	22.3 [c] ± 0.8	50
	21	11.3 ± 0.6	10.3 ± 0.3	20.3 [d] ± 0.9	55
FM 2 *Lactococcus lactis* subsp. *cremoris*, *Leuconostoc*, *Lactococcus lactis* subsp. *lactis*, and *Lactococcus lactis* subsp. *lactis diacetylactis*	0	12.2 ± 0.2	61.2 ± 3.1	71.1 [a] ± 3.4	-
	7	9.4 ± 0.4	25.2 ± 2.2	33.2 [b] ± 2.1	53
	14	12.3 ± 0.1	20.0 ± 2.4	31.2 [b] ± 2.3	56
	21	9.2 ± 0.2	15.3 ± 1.1	23.3 [c] ± 1.1	67
FM 3 *Lactobacillus delbrueckii* subsp. *bulgaricus*, *Lactobacillus delbrueckii* subsp. *lactis*, *Streptococcus thermophilus*, and *Bifidobacterium bifidum*	0	27.5 ± 2.3	84.0 ± 2.7	105.4 [a] ± 6.1	-
	7	17.8 ± 1.1	63.3 ± 0.9	77.2 [b] ± 3.6	27
	14	15.2 ± 0.8	34.8 ± 3.2	49.4 [c] ± 4.8	53
	21	11.3 ± 0.6	19.2 ± 1.8	30.3 [d] ± 2.6	71
FM 4 *Lactobacillus delbrueckii* subsp. *bulgaricus*, *Streptococcus thermophilus*, and *Bifidobacterium bifidum*	0	23.4 ± 0.2	41.1 ± 2.1	60.3 [b] ± 1.4	-
	7	16.2 ± 0.7	65.4 ± 5.2	79.2 [a] ± 6.3	+31
	14	13.3 ± 1.4	39.0 ± 3.3	51.1 [c] ± 4.1	15
	21	8.4 ± 0.6	23.1 ± 2.4	31.2 [d] ± 2.1	48

[1] The results are presented on the fresh weight as the mean of three replicates ± standard deviation. [2] Means in the column, for each fermented milk with the same letter, are not significantly different at $P < 0.05$.

Although yeasts are well known for their ability to produce folate during the fermentation process, the capacity of LAB for folate production is not yet clear. The reason is that both folate production and consumption are observed in culture media and in fermented foods [9,12]. Moreover, as noted by Saubade et al. [9], it is sometimes impossible to compare data on folate production by LAB in different studies due to the differences in methods applied and units in which results are expressed. In some studies, the authors measured the folate increase in the cell biomass, while other studies measured the folate increase in the supernatant of the culture medium. Furthermore, most studies in this field have tested the synthesis of folate by LAB in culture media. Although this is helpful in understanding the influence of different parameters, it may not be suitable for selecting LAB strains that increase folate amounts in real food matrices [9]. Table 2 shows the folate amounts determined by other authors and in own studies in milk fermented with LAB, by single strains or in combination.

Table 2. Folate content determined in fermented milk inoculated with LAB.

Folate Content [1] in Fermented Milk	Coverage of the Daily Demand for Folate (%) [2]	Producing Microorganisms	Ref.
5–50 μg/kg	1–12	*S. thermophilus*	[50]
20–50 μg/L	5–12	*S. thermophilus*	[55]
50–200 μg/kg	12–50	*S. thermophilus*	[56]
250–280 μg/L	63–70	*L. amylovorus*, *S. thermophilus* and *L. delbruecki* subsp. *bulgaricus*	[46]
10–70 μg/L	3–18	*L. delbruecki* subsp. *bulgaricus*	[57]
80–180 μg/L	20–45	*L. delbruecki* subsp. *bulgaricus*, *S. thermophilus*	[58]
27–35 μg/kg	7–9	*L. delbruecki* subsp. *bulgaricus*, *S. thermophilus*	[47]

Table 2. *Cont.*

Folate Content [1] in Fermented Milk	Coverage of the Daily Demand for Folate (%) [2]	Producing Microorganisms	Ref.
45 μg/kg	11	*L. delbruecki* subsp. *bulgaricus*, *S. thermophilus*	Own study
50–100 μg/L	13–25	*B. longum*	[16]
60–90 μg/kg	15–23	*S. thermophilus*, *B. animalis*	[17]
30–60 μg/kg	8–15	*S. thermophilus*, *B. longum*	[49]
60	15	*L. delbrueckii* subsp. *bulgaricus*, *S. thermophilus*, and *B. bifidum*	Own study
105 μg/kg	26	*L. delbruecki* subsp. *bulgaricus*, *L. lactis*, *S. thermophilus*, and *B. bifidum*	Own study
120–130 μg/kg	30–33	*Lc. lactis* subsp. *lactis*, *Lc. lactis* subsp. *cremoris*	[59]
2–20 μg/L	1–5	*Lc. lactis* subsp. *cremoris*	[53]
71 μg/kg	18	*Lc. lactis* subsp. *cremoris*, *Leuconostoc*, *Lc. lactis* subsp. *lactis*, and *L. lactis* subsp. *lactis diacetylactis*	Own study

[1] Folate content expressed in fresh weight basis. [2] Coverage of the daily demand for folate by adults, based on RDA (recommended daily allowance) of 400 μg set for men and women above 19 years old, by eating 1 kg or 1 L of the fermented milk product [60].

The authors emphasize that folate synthesis by LAB depends on many factors, including the strain, cultivation conditions, incubation time, the composition of the culture medium, and the presence of folate precursors or prebiotic supplementation [9, 14]. Nevertheless, depending on applied starter cultures and storage conditions, the folate concentration in fermented milk can be increased to more than 150 μg/L [46, 58]. In addition, as shown in Table 2, the use of a combination of different LAB may be more efficient than the use of single cultures [9]. Crittenden et al. [17] found that higher folate amounts were observed in milk fermented with a co-culture of *Streptococcus thermophilus and Bifidobacterium animalis* than in single cultures. Laiño et al. [58] noted that among different combinations, a strain of *Lactobacillus delbruecki* subsp. *bulgaricus* and *Streptococcus thermophilus* gave the best results in milk. In turn, Ayad [59] showed that the combination of *Lactococcus lactis* subsp. *lactis* and *Lactococcus lactis* subsp. *cremoris* was the most effective in folate production in Domiati cheese. In the authors' own study, the highest results were obtained with the mix of *Lactobacillus* species and *Streptococcus thermophilus* and *Bifidobacterium bifidum*.

However, the folate content of fermented milk enriched by LAB, including *Bifidobacteria*, is generally still too low to significantly contribute to the folate requirements at a satisfactory level. Most of the values presented in Table 2 of the coverage of the daily demand for folate by eating fermented milk products are not realistic. Even in the case of the tested FM 3 (with the highest folate content of 105 μg/kg), up to one kilogram of the product would have to be eaten during the day to cover approximately 25% of the daily requirement for this vitamin. Although LAB has been reported to significantly increase folate content in fermented products, it is still not enough to cover the daily requirements for this vitamin. Therefore, other solutions are needed to increase the folate content in foods fermented with LAB [9]. One option might be the combination of LAB with yeasts, which were reported to be more efficient in folate production when compared to LAB [52,61]. Another option, suggested by Saubade et al. [9], is to use LAB strains able to produce folate vitamers (such as 5-CH_3-H_4folate) which are generally more stable at low pH, which is the main folate form in fermented dairy products.

2.2. Fatty Acid Composition

Changes in the content of each group of fatty acids during storage at 8 °C in fermented milks were observed (Table 3). In all analyzed samples, fresh and stored, saturated fatty acids (SFA) were dominant. The mean content of SFA in the analyzed fermented milk inoculated with *Lactobacillus delbrueckii* subsp. *bulgaricus* and *Streptococcus thermophilus* (FM 1) was the lowest in fresh products. The fermented milk analyzed after 7, 14, and 21 days of storage was characterized by a significantly higher ($P < 0.05$) content of these acids. In stored fermented milk inoculated with *Lactococcus lactis* subsp. *cremoris, Leuconostoc, Lactococcus lactis* subsp. *lactis* and *Lactococcus lactis* subsp. *lactis diacetylactis* (FM 2) and *Lactobacillus delbrueckii* subsp. *bulgaricus,* and *Lactobacillus delbrueckii* subsp. *lactis Streptococcus thermophilus* and *Bifidobacterium bifidum* (FM 3), the mean content of SFA fluctuated. In fermented milk produced with *Lactobacillus delbrueckii* subsp. *Bulgaricus, Streptococcus thermophilus,* and *Bifidobacterium bifidum* (FM 4), the highest content of these acids was found in fresh products. Significantly lower ($P < 0.05$) SFA content was found in stored fermented milk. The content of PUFA (polyunsaturated fatty acids) in all analyzed products was at a similar level. In the case of MUFA (monounsaturated fatty acids) content, only in fermented milk inoculated with *Lactobacillus delbrueckii* subsp. *bulgaricus* and *Streptococcus thermophilus* (FM 1) and *Lactobacillus delbrueckii* subsp. *Bulgaricus, Streptococcus thermophilus,* and *Bifidobacterium bifidum* (FM 4) did not change significantly during storage. The contents of short-chain fatty acids (SCFA) in fermented milk inoculated with *Lactobacillus delbrueckii* subsp. *Bulgaricus, Streptococcus thermophilus,* and *Bifidobacterium bifidum* (FM 4) fluctuated during the storage period. In other analyzed fermented milks, the content of these acids changed significantly ($P < 0.05$) (Table 3).

Table 3. The content of fatty acid groups in fresh and refrigerated fermented milks (mg/g fat).

Fermented Milk (FM)	Fatty Acids	Days of Storage at 8 °C			
		0	7	14	21
FM 1 *Lactobacillus delbrueckii* subsp. *bulgaricus* and *Streptococcus thermophilus*	SCFA [1]	98.58 ± 8.34 ᵃ	74.76 ± 4.16 ᵇ	68.97 ± 7.79 ᵇ	94.02 ± 9.97 ᵃ
	SFA	481.28 ± 17.06 ᵇ	541.03 ± 20.45 ᵃ	546.80 ± 22.96 ᵃ	562.27 ± 19.44 ᵃ
	MUFA	58.33 ± 5.39 ᵃ	53.81 ± 2.73 ᵃ	56.48 ± 5.06 ᵃ	58.30 ± 8.41 ᵃ
	PUFA	21.78 ± 2.40 ᵃ [2]	20.26 ± 2.44 ᵃ	21.29 ± 2.63 ᵃ	22.27 ± 3.05 ᵃ
	n-3	2.05 ± 0.01 ᵃ	2.03 ± 0.11 ᵃ	2.14 ± 0.18 ᵃ	2.37 ± 0.39 ᵃ
	n-6	15.21 ± 0.27 ᵃ	15.25 ± 0.63 ᵃ	15.93 ± 0.88 ᵃ	16.76 ± 2.60 ᵃ
	n-6/*n*-3	7.41 ± 0.14 ᵃ	7.53 ± 0.12 ᵃ	7.47 ± 0.25 ᵃ	7.07 ± 0.11 ᵃ
	CLA	3.51 ± 0.22 ᵃ	3.38 ± 0.11 ᵃ	3.59 ± 0.31 ᵃ	3.73 ± 0.67 ᵃ
FM 2 *Lactococcus lactis* subsp. *cremoris, Leuconostoc, Lactococcus lactis* subsp. *lactis,* and *Lactococcus lactis* subsp. *lactis diacetylactis*	SCFA	99.92 ± 2.92 ᵃ	96.31 ± 13.94 ᵃ	73.51 ± 7.24 ᵇ	83.12 ± 5.47 ᵇ
	SFA	489.77 ± 9.15 ᵃ	496.57 ± 25.64 ᵃ	477.98 ± 20.01 ᵃ	502.29 ± 8.33 ᵃ
	MUFA	52.44 ± 3.21 ᶜ	58.13 ± 3.03 ᵇ	53.40 ± 3.88 ᵇᶜ	63.49 ± 3.71 ᵃ
	PUFA	20.17 ± 2.58 ᵃ	19.03 ± 2.03 ᵃ	18.70 ± 2.17 ᵃ	20.73 ± 3.95 ᵃ
	n-3	2.03 ± 0.03 ᵃ	2.05 ± 0.03 ᵃ	2.06 ± 0.02 ᵃ	2.06 ± 0.03 ᵃ
	n-6	13.97 ± 0.60 ᵃ	12.97 ± 0.47 ᵃ	12.90 ± 0.49 ᵃ	13.86 ± 1.01 ᵃ
	n-6/*n*-3	6.87 ± 0.39 ᵃ	6.32 ± 0.14 ᵇ	6.25 ± 0.30 ᵇ	6.71 ± 0.40 ᵃᵇ
	CLA	3.61 ± 0.17 ᵃᵇ	3.84 ± 0.16 ᵃ	3.43 ± 0.38 ᵇ	3.93 ± 0.09 ᵃ
FM 3 *Lactobacillus delbrueckii* subsp. *bulgaricus, Lactobacillus delbrueckii* subsp. *lactis, Streptococcus thermophilus,* and *Bifidobacterium bifidum*	SCFA	67.96 ± 5.52 ᶜ	65.32 ± 6.94 ᶜ	78.14 ± 4.27 ᵇ	88.37 ± 3.49 ᵃ
	SFA	607.54 ± 20.84 ᵃ	468.87 ± 12.05 ᵇ	478.63 ± 9.23 ᵇ	612.00 ± 25.36 ᵃ
	MUFA	52.15 ± 13.09 ᵃᵇ	43.24 ± 4.99 ᵇ	48.79 ± 2.76 ᵇ	62.75 ± 2.50 ᵃ
	PUFA	18.02 ± 4.60 ᵃ	16.59 ± 1.40 ᵃ	17.63 ± 2.85 ᵃ	17.83 ± 2.00 ᵃ
	n-3	1.86 ± 0.57 ᵃ	1.65 ± 0.22 ᵃ	1.80 ± 0.11 ᵃ	1.68 ± 0.12 ᵃ
	n-6	13.70 ± 3.98 ᵃ	12.29 ± 1.11 ᵃ	13.35 ± 1.04 ᵃ	12.76 ± 0.46 ᵃ
	n-6/*n*-3	7.39 ± 0.16 ᵃ	7.49 ± 0.39 ᵃ	7.43 ± 0.32 ᵃ	7.63 ± 0.78 ᵃ
	CLA	2.54 ± 0.28 ᶜ	2.71 ± 0.30 ᵇᶜ	3.06 ± 0.18 ᵃᵇ	3.42 ± 0.23 ᵃ
FM 4 *Lactobacillus delbrueckii* subsp. *bulgaricus, Streptococcus thermophilus,* and *Bifidobacterium bifidum*	SCFA	70.46 ± 15.24 ᵃ	64.76 ± 4.16 ᵃ	74.27 ± 7.45 ᵃ	75.02 ± 4.19 ᵃ
	SFA	570.24 ± 9.96 ᵃ	538.53 ± 22.74 ᵃ	563.05 ± 22.10 ᵃ	567.52 ± 24.15 ᵃ
	MUFA	52.15 ± 16.95 ᵃ	53.81 ± 2.73 ᵃ	56.48 ± 5.06 ᵃ	58.30 ± 8.41 ᵃ
	PUFA	18.07 ± 4.58 ᵃ	20.26 ± 2.44 ᵃ	20.04 ± 2.67 ᵃ	20.82 ± 2.08 ᵃ
	n-3	1.86 ± 0.53 ᵃ	2.03 ± 0.11 ᵃ	2.14 ± 0.18 ᵃ	2.24 ± 0.13 ᵃ
	n-6	13.59 ± 3.93 ᵃ	15.25 ± 0.63 ᵃ	15.68 ± 0.69 ᵃ	15.44 ± 0.65 ᵃ
	n-6/*n*-3	7.33 ± 0.15 ᵃᵇ	7.53 ± 0.12 ᵃ	7.36 ± 0.46 ᵃ	6.89 ± 0.28 ᵇ
	CLA	2.88 ± 0.65 ᵇ	3.38 ± 0.11 ᵃᵇ	3.59 ± 0.31 ᵃᵇ	3.89 ± 0.65 ᵃ

[1] SCFA—short-chain fatty acids (C4–C10); SFA—saturated fatty acids; MUFA—monounsaturated fatty acids; PUFA—polyunsaturated fatty acids, CLA (*cis*9*trans*11 C18:2)—conjugated linoleic acid. [2] Mean values, for fresh and stored fermented milk, in rows with the same letter are not significantly different at $P < 0.05$.

n-3 and n-6 polyunsaturated fatty acids are essential nutrients that cannot be synthesized in the body and must be obtained from the diet. It is important for the content of these acids to remain stable or increase during storage conditions to provide their beneficial effects. The results presented in Table 3 indicate that the cold storage time did not cause significant changes in the content of n-3 and n-6 acids. Only fluctuations in the content of these acids in stored fermented milks were observed. However, in a previous study [62], a significant decrease in n-3 and n-6 polyunsaturated fatty acid contents in cow milk yogurts produced with a starter culture containing *Streptococcus thermophilus* and *Lactobacillus delbrueckii* subsp. *bulgaricus* was observed during storage.

Research has shown that in stored fermented milk produced with the combination of *Lactococcus lactis* subsp. *cremoris, Leuconostoc, Lactococcus lactis* subsp. *lactis* and *Lactococcus lactis* subsp. *lactis diacetylactis* (FM 2) and *Lactobacillus delbrueckii* subsp. *bulgaricus, Streptococcus themophilus,* and *Bifidobacterium bifidum* (FM 4), significant changes ($P < 0.05$) in the n-6/n-3 acid ratio were observed. In other analyzed fermented milks, the ratio of these acids fluctuated slightly (Table 3). The proportions of specific groups of fatty acids in products are of special importance from a nutritional perspective. Excessive amounts of n-6 polyunsaturated fatty acids (PUFA) and a very high n-6/n-3 ratio promote the pathogenesis of many diseases, whereas increased levels of n-3 PUFA (a low n-6/n-3 ratio) exert suppressive effects [63,64].

The CLA content in fresh fermented milk ranged from 2.54 mg/g fat in FM 1 inoculated with *Lactobacillus delbrueckii* subsp. *bulgaricus* and *Streptococcus thermophilus* to 3.61 mg/g fat in FM2 produced with the combination of *Lactococcus lactis* subsp. *cremoris, Leuconostoc, Lactococcus lactis* subsp. *lactis,* and *Lactococcus lactis* subsp. *lactis diacetylactis* (Table 3). In all tested fermented milks, the time of refrigerated storage caused changes in CLA content, and the highest levels of this compound were found in all fermented milks stored for 21 days at 8 °C, with the highest amount of 3.93 mg/g fat in FM 2. However, in FM 1 and FM 2, the changes in *cis9trans11* C18:2 acid content after 7, 14, and 21 days were not significantly different ($P < 0.05$) compared to the CLA content in fresh fermented samples. Instead, in fermented milks enriched with *Bifidobacterium bifidum,* a significant increase in CLA contents was observed.

In FM 3 (with a mixture of *Lactobacillus delbrueckii* subsp. *bulgaricus, Lactobacillus delbrueckii* subsp. *lactis, Streptococcus thermophilus,* and *Bifidobacterium bifidum*), a significant increase was noted after 14 and 21 days of cold storage, while in FM 4 (inoculated with *Lactobacillus delbrueckii* subsp. *bulgaricus, Streptococcus thermophilus,* and *Bifidobacterium bifidum*) it was noted after 21 days. The obtained results demonstrate the potential of commercially available vaccines of LAB cultures, with special emphasis on *Bifidobacterium* to synthesize CLA during refrigerated storage. Research by other authors confirmed that the type of applied starter culture and storage time affects the content of CLA in fermented milk. According to a study by Domagała et al. [42], one of the seven starter cultures used by these authors (a yogurt culture ABY-2) caused an increase in the CLA content in stored fermented cream. Changes in CLA content in yogurts produced from cow's milk stored for 14 days at 5 °C were also demonstrated by Serafeimidow et al. [65]. According to their research, after seven days of storage, the yogurts from cow's milk were characterized by a higher content of CLA than the products analyzed on day 1. Significantly lower content of this acid was found by these authors in yogurts analyzed after 14 days of storage. Changes in CLA content in ecological and conventional fermented milk stored for seven days at 4 °C were also reported by Florence et al. [66]. According to Paszczyk et al. [67], out of three starter cultures used in the study, only one culture, Ceska-star Y508 (*Lactobacillus delbrueckii* subsp. *bulgaricus* and *Streptococcus thermophiles*), caused a significant increase in CLA content in the stored fermented milk drinks.

3. Materials and Methods

3.1. Samples

The research material was fermented milk (FM) from four productions. The FM covered by the study was produced using selected starter cultures. Four batches of fermented milk were produced using different selected starter cultures. Analyses were carried out for freshly produced FM samples and for samples stored at 8 ± 1 °C for 7, 14, and 21 days. Fermented milk was produced with the thermostat method according to the following technological scheme: raw milk was heated to 45 °C, centrifuged and degassed (80 kPa; 60 °C), then subjected to HTST pasteurization (72 °C/15 s; ALFA-LAVAL P20-HB pasteurizer, Lund, Sweden) and cooled to 6 °C. Afterward, it was normalized to a fat content of 2 ± 0.1% (addition of skim milk) and subjected to two-stage homogenization (18/5 MPa, 65 °C; homogenizer CN003, Spomasz Bełżyce, Poland) and long-term VHT pasteurization (90 °C/5 min; ALFA-LAVAL P20-HB pasteurizer, Lund, Sweden). After cooling to 45 °C, the milk was inoculated with pre-incubated for 2 h in 45 °C four different starter cultures (powder form) in the amount of 1 mL/L of milk. FM 1 was inoculated with FD-DVS YC-380 Yo-Flex containing *Lactobacillus delbrueckii* subsp. *bulgaricus* and *Streptococcus thermophilus* (Chr. Hansen, Hørsholm, Denmark). FM 2 was inoculated with FD-DVS FLORA DANICA starter culture containing *Lactococcus lactis* subsp. *cremoris*, *Leuconostoc*, *Lactococcus lactis* subsp. *lactis*, and *Lactococcus lactis* subsp. *lactis diacetylactis* (Chr. Hansen, Hørsholm, Denmark); FM 3 was inoculated with FD-DVS YC-180 Yo-Flex and BB-12 starter culture containing *Lactobacillus delbrueckii* subsp. *bulgaricus*, *Lactobacillus delbrueckii* subsp. *lactis*, *Streptococcus thermophilus*, and *Bifidobacterium bifidum*. FM 4 was inoculated with YC-X16 and BB-12 starter culture containing *Lactobacillus delbrueckii* subsp. *bulgaricus*, *Streptococcus thermophilus*, and *Bifidobacterium bifidum* (Chr. Hansen, Hørsholm, Denmark). The fermented milk drinks produced with the addition of selected starter cultures were transferred to unitary packages and left to ripen in thermostats (Binder GF115, Tuttlingen, Germany) at 43.5 °C until they reached pH 4.6 (c.a. 5 h).

3.2. Folate Analysis

3.2.1. Chemicals, Enzymes, and Standards

Water was purified in the Mili-Q system (Millipore; Vienna, Austria), acetonitrile was of HPLC grade, and the other chemicals were of analytical grade. Protease (E.C.3.4.24.31), obtained from Sigma Aldrich (St. Louis, MO, USA), was dissolved in 0.1 M phosphate buffer, pH 7.0, with 1% (*w*/*v*) sodium ascorbate and 0.1% (*v*/*v*) 2-mercaptoethanol (in the amount of 4 mg/mL) just before the analysis to avoid bacterial contamination, which can synthesize folate during incubation. Fresh rat plasma, used as a folate conjugase source, was purchased from Europa Bioproducts Ltd. (Cambridge, Great Britain) and prepared according to Patring et al. [68].

Folate standards: folic acid, 5-methyltetrahydrofolate (5-CH_3-H_4folate), 5-formyltetrahydrofolate (5-HCO-H_4folate), and tetrahydrofolate (H_4folate) were obtained from Sigma Aldrich (St. Louis, MO, USA). 10-formyl folic acid (10-HCO-folic acid) and 5,10-methenyltetrahydrofolate (5,10-CH^+-H_4folate) were obtained from Schircks Laboratories (Jona, Switzerland). All standards were prepared as described by Konings [69]. 10-formyldihydrofolate (10-HCO-H_2folate) was obtained from 5,10-CH^+-H_4folate according to Pfeiffer et al. [70].

3.2.2. Sample Preparation

The content of folate vitamers was analyzed in triplicate using the sample pretreatment method described by Gujska et al. [47]. During sample preparation, folates were protected against oxidation by carrying out the analysis under dim light and cooling the samples in ice after heating. Briefly, 10 g (accurate to 0.001 g) of fermented milk sample was inserted into a 30 mL PPCO Oak Ridge PPCO centrifuge tube (Nalgene; Rochester, NY, USA). Following this, 15 mL of an extraction buffer (0.1 M phosphate buffer, pH 7.0, with 1% (*w*/*v*) sodium ascorbate and 0.1% (*v*/*v*) 2-mercaptoethanol) were added. Samples were shaken (2500 rpm/10 s IKA Vortex 4 basic; Staufen, Germany) for 1 min and then transferred into a boiling water bath, heated for 15 min, shaken three times, and then cooled in ice. A total of 1 mL of the protease solution (4 mg/mL) and 0.25 mL of rat plasma conjugase were then added to each sample, and the samples were incubated at 37 °C for 4 h (POL-EKO; Rybnik, Poland). During incubation, samples were subjected to mild stirring (using a magnetic stirrer). Following this, they were heated in a boiling water bath for 5 min to inactivate enzymes, then cooled in ice and then centrifuged twice at 12,000 rpm/4 °C/20 min (MPW-350R; Warsaw, Poland). Each time, supernatants were collected in 50 mL amber volumetric flasks, which were filled up with the extraction buffer. The extract was filtered through the filter paper into amber glass bottles, flushed with nitrogen, and stored at −70 °C until the HPLC analysis.

Prior to the HPLC analysis, the samples were purified using Solid Phase Extraction (SPE) on Strong Anion Exchange (SAX) Bakerbond SPE JT cartridges (3 mL × 500 mg Solid Phase Extraction Column, PP (polypropylene), Quaternary Amine (N^+) Anion Exchange; Philipsburg, MT, USA) as described by Jastrebova et al. [71].

3.2.3. Folate Quantification

The chromatographic separation of folates was carried out according to Czarnowska and Gujska [72] using the HPLC system (Shimadzu Series LC-10A, Kyoto, Japan) and the C18 LC column: Synergi 4u Hydro-RP 80 Å (250 × 4.6 mm^2, 4 µm; Phenomenex; Torrance, CA, USA). The total separation time was 41 min. The chromatographic conditions for gradient elution were as follows: flow rate: 1 mL/min, volume injection 50 mL, column temperature 25 °C, UV detection: 290 nm; fluorescence detection: 290 nm excitation and 360 nm emission for 5-CH_3-H_4folate, 5-HCO-H_4folate, and H_4folate; 360 nm excitation and 460 nm emission for 10-HCO-folic acid. The mobile phase was acetonitrile with a 30 mM phosphoric acid buffer (pH 2.3). The gradient started at 5% acetonitrile and remained at that level for the first 8 min before being raised to 17.5% within 17 min. Peaks were identified based on the retention times of standards. Quantification of the identified individual folate vitamers was based on fluorescence detection using the external multilevel ($n = 8$) calibration curves. The results of folate vitamer content determination in the tested fermented milk were based on the fresh weight and presented as means with standard deviations from triplicates. The total folate content was the sum of 5-CH_3-H_4folate and H_4folate contents expressed as folic acid content using the molar absorption coefficient given by Blakely [73]. Differences in the mean total folate content in fresh and stored fermented milk samples were compared using the Duncan multiple range test, with a significance level of $P < 0.05$. The statistical analysis was carried out using Statistica software, version 13.1 (StatSoft; Cracow, Poland) [74].

3.3. *Fatty Acid Composition*

3.3.1. Fat Extraction

Fat was isolated from the analyzed fermented milks with the method of Folch et al. [75] with some modifications. Briefly, yogurts were heated to the temperature of 20 °C and thoroughly mixed. Approximately 10 g of samples (0.01 g) were homogenized (IKA Ultra-Turrax® T18 digital, Staufen, Germany) for 1 min with 100 mL of methanol. Next, 100 mL chloroform was added and homogenized for 2 min. The prepared mixture was filtered to a 500 mL glass cylinder. The solid residue was mixed in 200 mL chloroform: methanol (2:1 *v*/*v*) and homogenized again for 3 min. The mixture was transferred to the same cylinder. Then, 0.88% sodium chloride in water was added to the total filtrate (in the amount constituting 1/4 volume of filtrate); it was shaken vigorously for 1 min and left overnight to separate the layers. Next, the upper layer was removed using a water pump, and the lower layer was washed twice with a water–methanol mixture (1:1 *v*/*v*) and was filtered through anhydrous (VI) sodium sulfate. The solvent was evaporated. Methyl esters were prepared from the separated fat.

3.3.2. Preparation of Fatty Acid Methyl Esters

Fatty acid methyl esters were prepared according to the IDF method using a methanolic solution of KOH (ISO 15884:2002) [76]. N-hexane and 2 M KOH in methanol were added to each fat sample, and the mixture was then shaken. Sodium hydrogen sulfate ($NaHSO_4$) was then added, and the mixture was centrifuged (3000 rpm). The top layer of prepared methyl esters was collected for chromatographic analysis.

3.3.3. Gas Chromatography (GC) Analysis

The fatty acid contents were determined with the GC-FID method using the following: a capillary column (100 m × 0.25 mm i.d., film thickness 0.20 μm) (Chrompack, Middelburg, the Netherlands) with a CP Sil 88 stationary phase, and helium applied as a carrier gas at the flow rate of 1.5 mL/min. Sample injection volume was 0.4 μL (split: 50:1). Determinations were carried out under the following conditions: column temp. 60 °C (1 min)–180 °C, $\Delta t = 5$ °C/min, detector and injection temperatures of 250 °C and 225 °C, respectively.

3.3.4. Identification and Calculation of Fatty Acids

Identification of fatty acids was carried out based on a comparison of their retention time with the retention time of methyl esters of fatty acids of the reference milk fat (BCR Reference Materials) of the CRM 164 symbol and literature data [76–79]. The *cis*9*trans*11 C18:2 (CLA) isomer was identified using a mixture of CLA methyl esters (Sigma-Aldrich, Germany). The contents of fatty acids were calculated in mg/g fat according to the applicable standard (methyl ester of C21:0 acid, Sigma-Aldrich, Germany). The statistical analysis of results was carried out using Statistica version 13.1 (StatSoft; Cracow, Poland) [74] software based on a one-way analysis of variance (ANOVA) at a significance level at $p < 0.05$. The differences between mean values were evaluated using Duncan's test.

4. Conclusions

Many studies have presented promising results for the use of lactic acid bacteria to synthesize folate during fermentation. However, in other studies (including the authors' own study), either folate consumption by LAB or low folate production was observed. Commercial yogurts eaten in a normal daily portion cannot meet 10–20% of the daily recommended intake.

Combining LAB with different abilities to improve nutritional food quality is an option to maximize beneficial properties. Bio-fortified fermented milks offer a good alternative to develop a functional fermented food with increased amounts of essential compounds. There is an emerging opportunity for the food industry to use selected strains as starter cultures able to synthesize compounds such as vitamins and conjugated linoleic acid with a number of pro-health properties. For this reason, further research should focus on careful testing and selecting LAB strains able to produce folates, *cis*9*trans*11 C18:2 acid in high amounts, and keeping these nutrients stable during storage. Moreover, research is also required on the design and optimization of favorable conditions for such production.

Author Contributions: M.C.-K. performed the experiment, methodology, data analysis, review and writing—original draft preparation, project administration, and funding acquisition; B.P. performed the experiment, methodology, data analysis, review, and writing—original draft preparation. All authors have read and agreed to the published version of the manuscript.

Funding: Project financially supported by the Minister of Education and Science under the program entitled "Regional Initiative of Excellence" for the years 2019–2022, Project No. 010/RID/2018/19, amount of funding 12.000.000 PLN, and University of Warmia and Mazury funds Theme No. 17.610.008-300.

Institutional Review Board Statement: Not applicable.

Informed Consent Statement: Not applicable.

Data Availability Statement: Data is contained within the article.

Acknowledgments: We would like to express special thanks to Waldemar Brandt from the Department of Dairy Science and Quality Management for help in preparing yogurts for research.

Conflicts of Interest: The authors declare no conflict of interest.

Sample Availability: Samples of the compounds are available from the authors.

References

1. Blancquaert, D.; De Steur, H.; Gellynck, X.; Van Der Straeten, D. Present and Future of Folate Biofortification of Crop Plants. *J. Exp. Bot.* **2014**, *65*, 895–906. [CrossRef]
2. Najgebauer-Lejko, D.; Liszka, K.; Tabaszewska, M.; Domagała, J. Probiotic Yoghurts with Sea Buckthorn, Elderberry, and Sloe Fruit Purees. *Molecules* **2021**, *26*, 2345. [CrossRef]
3. Vora, A.; Riga, A.; Dollimore, D.; Alexander, K.S. Thermal Stability of Folic Acid. *Thermochim. Acta* **2002**, *392–393*, 209–220. [CrossRef]
4. Shulpekova, Y.; Nechaev, V.; Kardasheva, S.; Sedova, A.; Kurbatova, A.; Bueverova, E.; Kopylov, A.; Malsagova, K.; Dlamini, J. Ivashkin, V. The Concept of Folic Acid in Health and Disease. *Molecules* **2021**, *26*, 3731. [CrossRef]
5. Rad, A.H.; Khosroushahi, A.Y.; Khalili, M.; Jafarzadeh, S. Folate Bio-Fortification of Yoghurt and Fermented Milk: A Review. *Dairy Sci. Technol.* **2016**, *96*, 427–441. [CrossRef]
6. Rampersaud, G.C.; Kauwell, G.P.; Bailey, L.B. Folate: A Key to Optimizing Health and Reducing Disease Risk in the Elderly. *J. Am. Coll. Nutr.* **2003**, *22*, 1–8. [CrossRef]
7. Wright, A.; Finglas, P.; Southon, S. Proposed Mandatory Fortification of the UK Diet with Folic Acid: Have Potential Risks been Underestimated? *Trends Food Sci. Technol.* **2001**, *12*, 313–321. [CrossRef]
8. Buttriss, J. Strategies Designed to Increase Awareness about Folates and Health, and to Increase Folate Intake: A Review. *Trends Food Sci. Technol.* **2005**, *16*, 246–252. [CrossRef]
9. Saubade, F.; Hemery, Y.; Guyot, J.-P.; Humblot, C. Lactic Acid Fermentation as a Tool for Increasing the Folate Content of Foods. *Crit. Rev. Food Sci. Nutr.* **2017**, *57*, 3894–3910. [CrossRef] [PubMed]
10. WHO. *Guideline: Daily Iron and Folic Acid Supplementation in Pregnant Women*; World Health Organization: Geneva, Switzerland, 2012. Available online: http://www.who.int/nutrition/publications/micronutrients/guidelines/daily_ifa_supp_pregnant women/en/ (accessed on 30 August 2021).
11. FAO/WHO. Vitamin and Mineral Requirements in Human Nutrition, Second Edition. 2005. Available online: http://www.who int/nutrition/publications/micronutrients/9241546123/en/ (accessed on 30 August 2021).
12. LeBlanc, J.G.; De Giori, G.S.; Smid, E.J.; Hugenholtz, J.; Sesma, F. Folate Production by Lactic Acid Bacteria and Other Food-Grade Microorganisms. *Comm. Curr. Res. Educ. Top. Trends. Appl. Microbiol.* **2007**, *1*, 329–339.
13. Aryana, K.J.; Olson, D.W. A 100-Year Review: Yogurt and Other Cultured Dairy Products. *J. Dairy Sci.* **2017**, *100*, 9987–10013. [CrossRef]
14. Divya, J.B.; Nampoothiri, K.M. Folate Fortification of Skim Milk by a Probiotic Lactococcus Lactis CM28 and Evaluation of Its Stability in Fermented Milk on Cold Storage. *J. Food Sci. Technol.* **2015**, *52*, 3513–3519. [CrossRef]
15. Sybesma, W.; Starrenburg, M.; Tijsseling, L.; Hoefnagel, M.H.N.; Hugenholtz, J. Effects of Cultivation Conditions on Folate Production by Lactic Acid Bacteria. *Appl. Environ. Microbiol.* **2003**, *69*, 4542–4548. [CrossRef] [PubMed]
16. Lin, M.; Young, C. Folate Levels in Cultures of Lactic Acid Bacteria. *Int. Dairy J.* **2000**, *10*, 409–413. [CrossRef]
17. Crittenden, R.; Martinez, N.; Playne, M. Synthesis and Utilisation of Folate by Yoghurt Starter Cultures and Probiotic Bacteria. *Int. J. Food Microbiol.* **2003**, *80*, 217–222. [CrossRef]
18. Chin, S.; Liu, W.; Storkson, J.; Ha, Y.; Pariza, M. Dietary Sources of Conjugated Dienoic Isomers of Linoleic Acid, a Newly Recognized Class of Anticarcinogens. *J. Food Compos. Anal.* **1992**, *5*, 185–197. [CrossRef]
19. Jiang, J.; Björck, L.; Fondén, R. Production of Conjugated Linoleic Acid by Dairy Starter Cultures. *J. Appl. Microbiol.* **1998**, *8*. 95–102. [CrossRef]
20. Lin, H.; Boylston, T.D.; Luedecke, L.O.; Shultz, T.D. Factors Affecting the Conjugated Linoleic Acid Content of Cheddar Cheese. *J. Agric. Food Chem.* **1998**, *46*, 801–807. [CrossRef]
21. AkalIn, A.S.; Tokusoglu, O. A Potential Anticarcinogenic Agent: Conjugated Linoleic Acid (CLA). *Pak. J. Nutr.* **2003**, *2*, 109–11 [CrossRef]
22. Parodi, P.W. Anti-Cancer Agents in Milkfat. *Aust. J. Dairy Technol.* **2003**, *58*, 114–118.
23. Aydin, R. Conjugated Linoleic Acid: Structure, Sources and Biological Properties. *Turk. J. Vet. Anim. Sci.* **2003**, *29*, 189–195.
24. Park, Y. Conjugated Linoleic Acid (CLA): Good or Bad Trans Fat? *J. Food Compos. Anal.* **2009**, *22*, 4–12. [CrossRef]
25. Kee, J.-I.; Ganesan, P.; Kwak, H.-S. Bioactive Conjugated Linoleic Acid (CLA) in Milk. *Food Sci. Anim. Resour.* **2010**, *30*, 879–88 [CrossRef]
26. Precht, D.; Molkentin, J. Effect of Feeding on Trans Positional Isomers of Octadecenoic Acid in Milk Fats. *Milchwissenschaft* **199** *52*, 564–568.

27. Collomb, M.; Schmid, A.; Sieber, R.; Wechsler, D.; Ryhänen, E.-L. Conjugated Linoleic Acids in Milk Fat: Variation and Physiological Effects. *Int. Dairy J.* **2006**, *16*, 1347–1361. [CrossRef]
28. Frelich, J.; Šlachta, M.; Hanuš, O.; Špička, J.; Samková, E.; Węglarz, A.; Zapletal, P. Seasonal Variation in Fatty Acid Composition of Cow Milk in Relation to the Feeding System. *Anim. Sci. Pap. Rep.* **2012**, *30*, 219–229.
29. Kelsey, J.; Corl, B.; Collier, R.; Bauman, D. The Effect of Breed, Parity, and Stage of Lactation on Conjugated Linoleic Acid (CLA) in Milk Fat from Dairy Cows. *J. Dairy Sci.* **2003**, *86*, 2588–2597. [CrossRef]
30. Chilliard, Y.; Ferlay, A.; Mansbridge, R.M.; Doreau, M. Ruminant Milk Fat Plasticity: Nutritional Control of Saturated, Polyunsaturated, Trans and Conjugated Fatty Acids. *Anim. Res.* **2000**, *49*, 181–205. [CrossRef]
31. Zegarska, Z.; Paszczyk, B.; Rafałowski, R.; Borejszo, Z. Annual Changes in the Content of Unsaturated Fatty Acids with 18 Carbon Atoms, Including Cis9trans11 C18:2 (CLA) Acid, in Milk Fat. *Pol. J. Food Nutr. Sci.* **2006**, *56*, 409–414.
32. Hanuš, O.; Křížová, L.; Samková, E.; Špička, J.; Kučera, J.; Klimešová, M.; Roubal, P.; Jedelská, R. The Effect of Cattle Breed, Season and Type of Diet on the Fatty Acid Profile of Raw Milk. *Arch. Anim. Breed.* **2016**, *59*, 373–380. [CrossRef]
33. Shantha, N.C.; Decker, E.A.; Ustunol, Z. Conjugated Linoleic Acid Concentration in Processed Cheese. *J. Am. Oil Chem. Soc.* **1992**, *69*, 425–428. [CrossRef]
34. Bisig, W.; Eberhard, P.; Collomb, M.; Rehberger, B. Influence of Processing on the Fatty Acid Composition and the Content of Conjugated Linoleic Acid in Organic and Conventional Dairy Products—A Review. *Le Lait* **2007**, *87*, 1–19. [CrossRef]
35. Salamon, R.V.; Lóki, K.; Varga-Visi, E.; Mándoki, Z.; Csapó, J. Increase of Conjugated Linoleic Acid Content of Dairy Products by Adding Sunflower Oil. *Krmiva* **2009**, *51*, 99–103.
36. Junior, O.O.S.; Pedrao, M.R.; Dias, L.F.; Paula, L.N.; Coro, F.A.; De Souza, N.E. Fatty Acid Content of Bovine Milkfat from Raw Milk to Yoghurt. *Am. J. Appl. Sci.* **2012**, *9*, 1300–1306. [CrossRef]
37. Seçkin, A.K.; Gursoy, O.; Kinik, O.; Akbulut, N. Conjugated Linoleic Acid (CLA) Concentration, Fatty Acid Composition and Cholesterol Content of Some Turkish Dairy Products. *LWT* **2005**, *38*, 909–915. [CrossRef]
38. Prandini, A.; Sigolo, S.; Tansini, G.; Brogna, N.; Piva, G. Different Level of Conjugated Linoleic Acid (CLA) in Dairy Products from Italy. *J. Food Compos. Anal.* **2007**, *20*, 472–479. [CrossRef]
39. Alonso, L.; Cuesta, E.P.; Gilliland, S.E. Production of free conjugated linoleic acid by Lactobacillus acidophilus and Lactobacillus casei of human intestinal origin. *J. Dairy Sci.* **2003**, *86*, 1941–1946. [CrossRef]
40. Kim, Y.; Liu, R. Increase of Conjugated Linoleic Acid Content in Milk by Fermentation with Lactic Acid Bacteria. *J. Food Sci.* **2002**, *67*, 1731–1737. [CrossRef]
41. Sieber, R.; Collomb, M.; Aeschlimann, A.; Jelen, P.; Eyer, H. Impact of Microbial Cultures on Conjugated Linoleic Acid in Dairy Products—A Review. *Int. Dairy J.* **2004**, *14*, 1–15. [CrossRef]
42. Domagała, J.; Sady, M.; Najgebauer-Lejko, D.; Czernicka, M.; Wieteska, I. The Content of Conjugated Linoleic Acid (CLA) in Cream Fermented Using Different Starter Cultures. *Biotechnol. Anim. Husb.* **2009**, *25*, 745–751.
43. Hennessy, A.; Ross, R.; Devery, R.; Stanton, C. Optimization of a Reconstituted Skim Milk Based Medium for Enhanced CLA Production by Bifidobacteria. *J. Appl. Microbiol.* **2009**, *106*, 1315–1327. [CrossRef]
44. Ogawa, J.; Kishino, S.; Ando, A.; Sugimoto, S.; Mihara, K.; Shimizu, S. Production of Conjugated Fatty Acids by Lactic Acid Bacteria. *J. Biosci. Bioeng.* **2005**, *100*, 355–364. [CrossRef]
45. Hugenschmidt, S.; Schwenninger, S.M.; Gnehm, N.; Lacroix, C. Screening of a Natural Biodiversity of Lactic and Propionic Acid Bacteria for Folate and Vitamin B12 Production in Supplemented Whey Permeate. *Int. Dairy J.* **2010**, *20*, 852–857. [CrossRef]
46. Laiño, J.E.; del Valle, M.J.; de Giori, G.S.; LeBlanc, J.G.J. Applicability of a Lactobacillus Amylovorus Strain as Co-Culture for Natural Folate Bio-Enrichment of Fermented Milk. *Int. J. Food Microbiol.* **2014**, *191*, 10–16. [CrossRef]
47. Gujska, E.; Czarnowska, M.; Michalak, J. Content of Folates in Fresh and Cold Stored Kefirs and Yoghurts. *Zywnosc Nauka Technol. Jakosc/Food Sci. Technol. Qual.* **2014**, *5*, 124–133. [CrossRef]
48. Strandler, H.S.; Patring, J.; Jägerstad, M.; Jastrebova, J. Challenges in the Determination of Unsubstituted Food Folates: Impact of Stabilities and Conversions on Analytical Results. *J. Agric. Food Chem.* **2015**, *63*, 2367–2377. [CrossRef]
49. Holasová, M.; Fiedlerová, V.; Roubal, P.; Pechačová, M. Possibility of Increasing Natural Folate Content in Fermented Milk Products by Fermentation and Fruit Component Addition. *Czech J. Food Sci.* **2005**, *23*, 196–201. [CrossRef]
50. Holasova, M.; Fiedlerova, V.; Roubal, P.; Pechacova, M. Biosynthesis of Folates by Lactic Acid Bacteria and Propionibacteria in Fermented Milk. *Czech J. Food Sci.* **2004**, *22*, 175–181. [CrossRef]
51. Jagerstad, M.; Jastrebova, J.; Svensson, U. Folates in Fermented Vegetables—A Pilot Study. *LWT* **2004**, *37*, 603–611. [CrossRef]
52. Kariluoto, S.; Aittamaa, M.; Korhola, M.; Salovaara, H.; Vahteristo, L.; Piironen, V. Effects of Yeasts and Bacteria on the Levels of Folates in Rye Sourdoughs. *Int. J. Food Microbiol.* **2006**, *106*, 137–143. [CrossRef]
53. Gangadharan, D.; Nampoothiri, K.M.; Gangadharan, D.; Nampoothiri, K.M. Folate Production Using *Lactococcus lactis* ssp. *cremoris* with Implications for Fortification of Skim Milk and Fruit Juices. *LWT* **2011**, *44*, 1859–1864. [CrossRef]
54. Wigertz, K.; Jägerstad, M. Comparison of a HPLC and Radioprotein-Binding Assay for the Determination of Folates in Milk and Blood Samples. *Food Chem.* **1995**, *54*, 429–436. [CrossRef]
55. Iyer, R.; Tomar, S.K.; Mohanty, A.K.; Singh, P.; Singh, R. Bioprospecting of Strains of Streptococcus Thermophilus from Indian Fermented Milk Products for Folate Production. *Dairy Sci. Technol.* **2011**, *91*, 237–246. [CrossRef]

56. Padalino, M.; Perez-Conesa, D.; López-Nicolás, R.; Frontela-Saseta, C.; Ros, G. Effect of Fructooligosaccharides and Galactooligosaccharides on the Folate Production of some Folate-Producing Bacteria in Media Cultures or Milk. *Int. Dairy J.* **2012**, *27*, 27–33. [CrossRef]
57. Lin, M.Y.; Young, C.M. Biosynthesis of Folates by *Streptococcus thermophilus* and *Lactobacillus delbrueckii* ssp. *bulgaricus*. *J. Food Drug Anal.* **2000**, *8*, 195–199.
58. Laiño, J.E.; del Valle, M.J.; de Giori, G.S.; LeBlanc, J.G. Development of a High Folate Concentration Yogurt Naturally Bio-Enriched Using Selected Lactic Acid Bacteria. *LWT* **2013**, *54*, 1–5. [CrossRef]
59. Ayad, E.H. Starter Culture Development for Improving Safety and Quality of Domiati Cheese. *Food Microbiol.* **2009**, *26*, 533–541. [CrossRef] [PubMed]
60. Jarosz, M.; Stoś, K.; Przygoda, B.; Matczuk, E.; Stolińska-Fiedorowicz, H.; Kłys, W. Witaminy. In *Normy dla populacji Polski*; Jarosz, M., Ed.; IŻŻ: Warsaw, Poland, 2017; pp. 166–170. Available online: https://ncez.pl/upload/normy-net-1.pdf (accessed on 30 August 2021).
61. Korhola, M.; Hakonen, R.; Juuti, K.; Edelmann, M.; Kariluoto, S.; Nyström, L.; Sontag-Strohm, T.; Piironen, V. Production of Folate in Oat Bran Fermentation by Yeasts Isolated from Barley and Diverse Foods. *J. Appl. Microbiol.* **2014**, *117*, 679–689. [CrossRef]
62. Paszczyk, B.; Łuczyńska, J.; Polak-Śliwińska, M. The Effect of Storage on the Yogurt Fatty Acid Profile. *Mljekarstvo* **2020**, *70*, 59–70. [CrossRef]
63. Simopoulos, A.P. The Importance of the Omega-6/Omega-3 Fatty Acid Ratio in Cardiovascular Disease and Other Chronic Diseases. *Exp. Biol. Med.* **2008**, *233*, 674–688. [CrossRef]
64. Wijendran, V.; Hayes, K. Dietary N-6 and N-3 Fatty Acid Balance and Cardiovascular Health. *Annu. Rev. Nutr.* **2004**, *24*, 597–615. [CrossRef] [PubMed]
65. Serafeimidou, A.; Zlatanos, S.; Kritikos, G.; Tourianis, A. Change of Fatty Acid Profile, Including Conjugated Linoleic Acid (CLA) Content, during Refrigerated Storage of Yogurt Made of Cow and Sheep Milk. *J. Food Compos. Anal.* **2013**, *31*, 24–30. [CrossRef]
66. Florence, A.C.R.; Béal, C.; Silva, R.C.; Bogsan, C.S.; Pilleggi, A.L.O.; Gioielli, L.A.; Oliveira, M.N. Fatty Acid Profile, Trans-Octadecenoic, A-Linolenic and Conjugated Linoleic Acid Contents Differing in Certified Organic and Conventional Probiotic Fermented Milks. *Food Chem.* **2012**, *135*, 2207–2214. [CrossRef] [PubMed]
67. Paszczyk, B.; Brandt, W.; Łuczyńska, J. Isomers of c18:1 and c18:2 Acids in Fresh and Stored Fermented Milks Produced with Selected Starter Cultures. *Czech J. Food Sci.* **2016**, *34*, 391–396. [CrossRef]
68. Patring, J.; Wandel, M.; Jägerstad, M.; Frølich, W. Folate content of Norwegian and Swedish Flours and Bread Analysed by Use of Liquid Chromatography—Mass Spectrometry. *J. Food Compos. Anal.* **2009**, *22*, 649–656. [CrossRef]
69. Konings, E.J.M. A Validated Liquid Chromatographic Method for Determining Folates in Vegetables, Milk Powder, Liver, and Flour. *J. AOAC Int.* **1999**, *82*, 119–127. [CrossRef]
70. Pfeiffer, C.M.; Rogers, L.; Gregory, J.F. Determination of Folate in Cereal-Grain Food Products Using Trienzyme Extraction and Combined Affinity and Reversed-Phase Liquid Chromatography. *J. Agric. Food Chem.* **1997**, *45*, 407–413. [CrossRef]
71. Jastrebova, J.; Witthöft, C.; Grahn, A.; Svensson, U.; Jägerstad, M. HPLC Determination of Folates in Raw and Processed Beetroots. *Food Chem.* **2003**, *80*, 579–588. [CrossRef]
72. Czarnowska, M.; Gujska, E. Effect of Freezing Technology and Storage Conditions on Folate Content in Selected Vegetables. *Plant Foods Hum. Nutr.* **2012**, *67*, 401–406. [CrossRef]
73. Blakley, R.L. The Biochemistry of Folic Acid and Related Pteridines. In *North-Holland Research Monographs*; North-Holland Publishing Company: Amsterdam, The Netherlands, 1969; pp. 1–570.
74. *STATISTICA*, version 13.1; Statsoft: Kraków, Poland, 2013.
75. Christie, W.W. The Isolation of Lipids from Tissues. Recommended Procedures. Chloroform-Methanol (2:1, V/V) Extraction and "Folch" Wash. In *Lipid Analysis. Isolation, Separation, Identification and Structural Analysis of Lipids*; Christie, W.W., Ed.; Pergamon Press Oxford: New York, NY, USA; Toronto, Japan; Sydney, Australia; Braunschweig, Germany, 1973; pp. 39–40.
76. *Milkfat: Preparation of Fatty Acid Methyl Esters*; ISO 15884:2002 (IDF 182:2002); International Organization for Standardization: Geneva, Switzerland, 2002.
77. Roach, J.A.; Mossoba, M.M.; Yurawecz, M.; Kramer, J.K. Chromatographic Separation and Identification of Conjugated Linoleic Acid Isomers. *Anal. Chim. Acta* **2002**, *465*, 207–226. [CrossRef]
78. Kramer, J.K.G.; Cruz-Hernandez, C.; Deng, Z.; Zhou, J.; Jahreis, G.; Dugan, M.E.R. Analysis of Conjugated Linoleic Acid and Trans 18:1 Isomers in Synthetic and Animal Products. *Am. J. Clin. Nutr.* **2004**, *79*, 1137S–1145S. [CrossRef]
79. Ledoux, M.; Chardigny, J.-M.; Darbois, M.; Soustre, Y.; Sébédio, J.-L.; Laloux, L. Fatty Acid Composition of French Butters, with Special Emphasis on Conjugated Linoleic Acid (CLA) Isomers. *J. Food Compos. Anal.* **2005**, *18*, 409–425. [CrossRef]

molecules

MDPI

Review

Whipping Creams: Advances in Molecular Composition and Nutritional Chemistry

Khadija Florence Dabo [1,2], Christine Chèné [1], Anne-Laure Fameau [3,*] and Romdhane Karoui [2]

[1] Adrianor, 62217 Tilloy-Lès-Mofflaines, France; f.dabo@adrianor.com (K.F.D.); c.chene@adrianor.com (C.C.)
[2] University of Artois, University of Lille, University of Littoral Côte d'Opale, University of Picardie Jules Verne, University of Liège, INRAE, Junia, UMR-T 1158, BioEcoAgro, 62300 Lens, France; romdhane.karoui@univ-artois.fr
[3] CNRS, INRAE, Centrale Lille, UMET, University of Lille, 59000 Lille, France
* Correspondence: anne-laure.fameau@inrae.fr

Abstract: Whipping cream (WC) is an oil-in-water (O/W) emulsion used in food industry that can be transformed into aerated foam. The cream market has expanded significantly, driven by consumer demands for healthier and higher-quality products, leading to significant scientific research and innovation. This review focuses on formulation challenges related to ingredients such as fats, emulsifiers, and stabilizers, and how these components interact to form a stable emulsion and foam structure. Many studies have aimed to enhance the physicochemical, functional, and nutritional characteristics of WC by fine-tuning formulation parameters. A major focus was to address the health concerns linked to the high saturated fat content in milk fat (MF) by developing healthier alternatives. These include modifying the fat content, developing low-fat formulations, and introducing plant-based substitutes for dairy creams. The participation of additives to improve the properties of whipping cream was also investigated in many recent studies. The use of plant proteins, hydrocolloids, and emulsifiers has been explored, highlighting their effectiveness in enhancing emulsifying and foaming properties. This review summarizes recent advancements in whipping cream formulation, emphasizing the role of additives and alternative ingredients in meeting consumer preferences for healthier, more sustainable whipping cream products with enhanced functional, sensory, and nutritional properties.

Keywords: whipping cream; foam; proteins; fats; surfactant; plant-based

1. Introduction

Whipping cream is an oil-in-water emulsion widely used in the food industry, as-is or transformed into an aerated foam that can have pleasant sensory characteristics and a stable foamy texture when used for desserts [1]. The whipping cream market has seen significant growth, with its value expected to rise from 819.1 million in 2021 to 1268.5 million by 2029 according to the Data Bridge Market Research [2]. This has led to an increase in scientific research on whipping creams, exploring its functional properties [3–5], production process [6,7], and how to improve its nutritional properties [8,9]. In recent years, many companies have offered new and innovative products to meet the demands of consumers.

Historically, the production of whipping cream was achieved through the natural skimming of milk due to the inherent difference in density between the fat globules and the aqueous phase of the milk. Subsequently, numerous innovative technologies have been developed, including those based on skimming and centrifugation, as well as those involving the assembly or reconstitution of WC with dairy ingredients [10]. Whipping cream is commonly composed of water, semi-crystalline fat, sugars, protein, stabilizers, and low-molecular-weight emulsifiers [11]. The stability of this emulsion is crucial; if too stable, foam formation is hindered, while too unstable leads to flocculation and phase separation of emulsion [12,13]. The semi-crystalline state of fat plays a significant role in the stability of the foam structure. During the process of whipping, fat droplets collide with one another, and crystals in one droplet may pierce the interfacial layer of the adjacent droplet, leading to fat coalescence [14]. Early research focused on the partial coalescence of fat, essential for the formation of a stable foam structure [15–18]. Recent studies have expanded to include the quality characteristics of whipping cream, such as its physical, functional, rheological, and nutritional properties [3,19–21]. Consumers demand high-quality nutrition, healthy and environmentally friendly ingredients, and excellent sensory qualities, while traders seek better storage and whipping properties to reduce costs and increase commercial value [22].

The milk fat in whipping cream contains relatively large amounts of saturated fatty acids which contribute to the high intake of these nutrients and therefore increase the risk of obesity and cardiovascular disease [22]. Thus, the high-fat content poses health risks, prompting the development of low-fat alternatives. That is why the research has intensely focused on the development of low-fat formulation, and explored the effects of various additives (hydrocolloids, proteins, emulsifiers, and lipid compounds) on WC and the final nutritional properties of whipped cream [5,23,24]. Another recent trend in the whipping cream industry has been the development of plant-based alternatives to traditional whipping cream, catering to the growing demand for vegan and dairy-free options [4,25,26]. Incorporating plant-based components in whipping cream manufacturing mitigates the impact of seasonal fluctuations on MF outcomes. This enhances the product's functionality, nutritional value, and cost-effectiveness while optimizing quality and efficiency [27–29].

The use of additives is a necessity due to the complexity and instability of whipping cream; it also helps to enhance the functionality and quality of the product. Consequently, additives are instrumental in improving emulsification, foaming, sensory attributes, and rheological properties of WC. A variety of components, including hydrocolloids, proteins, lipid compounds, carbohydrates, and emulsifiers, are employed as additives, offering a range of benefits and drawbacks. For example, hydrocolloids such as xanthan gum (XG) have been demonstrated to improve foam stability and viscosity [30]. Similarly, proteins like whey protein and sodium caseinate (NaC) have been shown to enhance emulsification [31] and foam stability [32,33] under certain conditions (pH, hydrolysis, etc.). Emulsifiers, such as sucrose esters and triglycerol monostearate, improve overrun and stability by reducing interfacial tension and promoting the partial coalescence of fat globules [34]. Carbohydrates also play a crucial role in enhancing the whipping properties [4]. Sensory properties, including texture, color, and overall taste, are also critical for consumer acceptance.

Therefore, whipping cream is a popular product with a growing market, but it encounters obstacles pertaining to stability, fat content, and composition, as well as the necessity to enhance its quality and nutritional characteristics. The use of additives represents a pivotal aspect of confronting these challenges. This review examines the recent scientific advancements made over the past three years regarding the optimization of whipping cream composition and the enhancement of its functional and nutritional properties. The objective is to develop healthier and more appealing products that align with the needs of both consumers and industry.

2. Whipping Cream: Emulsion and Foam Characterization

2.1. Main Components of Whipping Cream and Their Key Roles

A whipping cream is defined as an emulsion that comprises three principal phases: the fatty phase, the aqueous phase, and the interface between the fatty and aqueous phase [35]. The aqueous phase typically consists of skimmed milk or water, while the fatty phase may include milk fat, vegetable fat, or a fat analog as an alternative. The third phase is the interface, composed of emulsifiers and/or stabilizers such as animal or vegetable proteins and low-molecular-weight surfactants (LMWS). Additionally, various other ingredients may be present in whipping cream recipes, including stabilizers (e.g., carrageenan) and sugars (e.g., lactose) [36]. Each of these ingredients plays a distinct role in the emulsion and the subsequent formation of foam.

Fat is a principal component of whipping cream due to its role in emulsification and foam stabilization [37]. The fat content in dairy whipping cream generally ranges from 30% to 40%, but in vegan or low-fat alternatives, this may be reduced to around 20% [5,23]. The fatty phase is combined with the soluble phase in the presence of a surface-active agent to facilitate the process of emulsification and to stabilize the fat droplets that are formed. The droplet size depends on the processing methods used (such as homogenization and fat maturation) and formulation factors, including fat composition and emulsifier properties [10]. The fat globule sizes are expected to be lower than 1 μm for industrial WC, since larger fat globules are known to increase the emulsion destabilization rate [38]. The fatty phase can derive from either dairy or plant sources. However, whatever the origin of the fat, the fat phase should exhibit low crystallization to facilitate the uniform formation of fat crystals during cooling and a high melting point to prevent crystallization/melting/recrystallization during storage, which could alter the functional properties [17,37]. The formation of crystals within fat droplets is a consequence of the maturation process, which involves the storage of the fat at a relatively low temperature (between 4 and 10 °C, depending on the specific recipe) for a minimum of 24 h following the production of whipping cream. The type of crystals formed is a function of multiple factors, including the origin of the fat, the type of emulsifier, and the temperature at which crystallization occurs. The formation of β or β' crystals is preferable to α crystals for whipping cream due to their stability [39]. Hydrogenated vegetable oil or native oils with a high saturated fat content, such as coconut or palm, are commonly employed as partial or total substitutes for milk fat [29]. When used alone or in combination with other vegetable oils, they exhibit a diverse range of crystallization and melting behaviors, which can impart novel properties to the cream in terms of functionality (stability under thermal shock, ability to expand, etc.) [10].

The aqueous phase is composed of water or skimmed milk, which are commonly used to adjust the fat content of dairy whipping creams and also serve as a continuous phase in emulsions where soluble components such as proteins and sugars are solubilized prior to emulsification. The soluble phase typically constitutes approximately 50 to 60% of the total recipe.

Milk proteins are naturally present in skimmed milk or native cream, where they act as surface agents. It is possible to incorporate supplementary milk or vegetal proteins into a whipping cream recipe at a concentration ranging from 0.5 and 2% [40]. In this product category, caseins are typically the most utilized, particularly sodium caseinate, even in plant-based WC [10]. In response to the growing demand for naturality, plant proteins derived from soy, chickpeas, and kidney beans are increasingly utilized as caseinate replacers [5,41]. The proteins are expected to adsorb rapidly at the interface and reduce interfacial tension, preventing the destabilization of emulsions and maintaining the stability of air bubbles [36].

Additionally, a variety of components are incorporated into whipping creams to optimize their functional, physicochemical, rheological, and sensory characteristics. These components include emulsifiers such as lecithin and stabilizers such as carrageenan. The principal function of these additives is to strengthen the emulsion and stabilize air bubbles, thus preventing phase separation or collapse. Emulsifiers have the capacity to displace or coexist with proteins at the interface, thereby influencing the formation of interfacial layers and the properties of whipped cream [42]. Emulsifiers contribute to the stabilization of the WC emulsion by adsorbing at the oil–water interface, which induces a reduction in interfacial tension and prevents the separation of fat. In the foaming process, the emulsifiers facilitate the formation of a more robust structure surrounding the air bubbles. Proteins with a high molecular weight can partially stabilize the fats. Therefore, it is essential to utilize LMWS to develop a more robust membrane that can support long-term emulsion storage [34]. Stabilizers are expected to form a cohesive, resistant monolayer with the objective of stabilizing the emulsion and foam. Carrageenan is typically found in whipping cream, but other stabilizers, including starch, gum, and pectin, can be employed as alternatives [14]. The addition of these stabilizers serves to enhance the structural integrity of the whipped cream by thickening the soluble phase, thereby providing a firmer structure that effectively prevents foam collapse and syneresis. Ingredients such as sugars slightly enhance the flavor and can also influence the physical properties of whipped cream. They can be directly added during production and during whipping. They behave as thickening agents to improve emulsion and foam stability [14].

All these components function collectively to facilitate a stable emulsion, thereby optimizing the production of the desired fluffy texture when whipped.

2.2. Whipping Cream: Mechanism of Formation and Stabilization

In whipping cream, the key phenomenon is the so-called partial coalescence required to obtain a stable semi-solid foam system with a desired texture after whipping. However, this phenomenon only needs to occur during whipping and partial coalescence should be avoided during the long-term storage of the whipping cream. During the process of whipping, the collision of fat globules containing solid fat results in the formation of partially coalesced fat globules, which subsequently form a membrane that covers air bubbles and stabilizes the whipped foam [7,34]. The created fat droplets induce the formation of a 3D network of interconnected semi-solid droplets, leading to an increase in the apparent viscosity of the product. When the solid fat content is less than or equal to 15%, the phenomenon is referred to as total coalescence. If the solid fat content is between 30 and 45%, this is called partial coalescence, and a solid fat content over 50% is considered fully coalesced [38].

As illustrated in Figure 1, the emulsion undergoes a three-stage transformation into foam during the whipping process, as described by Han et al. [6]. Initially, whipping causes the formation of large air bubbles within the system. Then, the protein present in the soluble phase adsorbs onto the surface of these bubbles, thereby stabilizing the initial foam. During this phase, fat droplets collide with each other, primarily due to shear-induced partial coalescence. Subsequently, under continuous shearing, these large air bubbles undergo a reduction in diameter, accompanied by an increase in the number of partially coalesced fat droplets. Ultimately, the partially coalesced fat droplet forms a network structure under the protein thin membrane that initially surrounds the bubbles, imparting stiffness to the whipped cream structure [6,14].

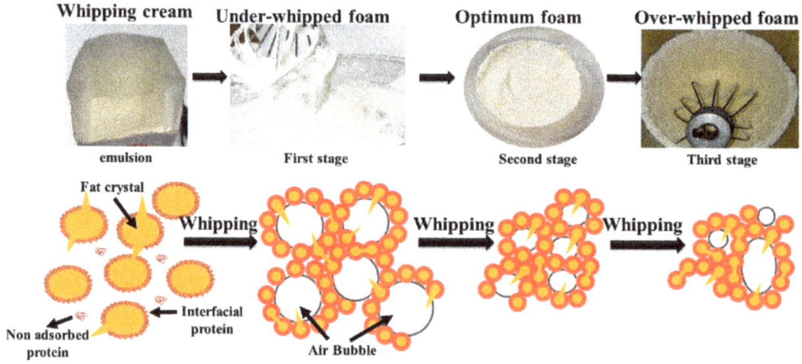

Figure 1. Schematic showing the mechanisms of coalescence and whipping inflation of fat globules in whipping cream depending on the whipping time.

The protein layer plays a crucial role in maintaining the stability of emulsions and foams. It accomplishes this by wrapping around the fat droplets and air bubbles, forming a protective barrier that prevents them from merging and collapsing [6]. In commercial preparations, stabilizers such as carrageenan or xanthan gum are frequently incorporated to improve the capacity of whipping cream to retain air and sustain its structure over time [14]. It is also imperative to regulate the temperature of whipping cream during the storage period and prior to whipping. It is recommended that dairy-based WC be stored at a low temperature to maintain the integrity of the fat crystals. Prior to usage, whipping cream with dairy or plant-based ingredients should be stored at a temperature between 0 °C and 4 °C to ensure foamability. Furthermore, the duration and speed of whipping are of consequence. If the cream is over-whipped, the fat droplets will coalesce entirely, which will result in the breakdown of the foam structure and a collapse rather than a stable foam [26].

An understanding of the composition of whipping cream and the role of each ingredient allows manufacturers to tailor the properties of whipped cream and meet the needs of consumers and commercial applications. This ensures both the organoleptic and functional properties of WC.

3. Effect of Emulsifiers at the Interface

Fat crystallization is crucial to the formation and stabilization of whipped cream, and the variability of the crystal structures affect the partially coalesced network formed during whipping, which influences whipped creams' stability [43]. Milk fat with appropriate melting properties and a high saturated fatty acid content is commonly used to obtain optimal functionality even if it presents health risks. Using anhydrous milk fat (AMF) is a way to decrease the possible impact of issues pertaining to fat origin (seasonality, type of cow, cow feed, etc.). However, AMF's characteristics are influenced by various factors, making standardization challenging. Whipping cream that is non-dairy based is traditionally made using hydrogenated palm kernel oil (HPKO), which contains trans-fatty acids and saturated fats. Depending on their composition, fats have different crystallization profiles, which impact their emulsifying and foaming properties. Blending AMF with HPKO is a practical solution, as HPKO has a high melting point and a high medium and long-chain saturated fatty acid content, contributing to desirable melt-in-the-mouth characteristics [29]. The compatibility of AMF and HPKO, which both crystallize in the β′-form, makes this fat blend ideal for whipped cream. However, challenges such as flocculation and coalescence during storage necessitate the comprehensive regulation of fat crystallization and interfacial properties. Moreover, hydrogenated palm kernel oil contains trans fatty acids and saturated fats, which pose health risks. To mitigate these risks, fat substitutes using highly unsaturated liquid vegetable oils are being developed. However, substituting liquid vegetable oils for solid fats in whipped cream destabilizes its structure, leading to low firmness but increases foam stability. Emulsifiers play a crucial role in the whipping and foaming process by inducing fat globule destabilization and promoting partial coalescence. Emulsifiers can reduce protein adsorption at oil–water interfaces through the 'orogenic displacement' mechanism. This ability of surfactants to displace proteins varies according to their chemical structure, hydrophilic–lipophilic balance (HLB), and interactions with bulk fats. However, formulation rules are still not clear in the literature. That is why recent studies focus on the determination of the relationship between the surfactant nature, HLB, and the ability to replace proteins at the interface, facilitating the penetration of fat crystals during whipping and perhaps contributing to partial coalescence.

3.1. *Low-Molecular-Weight Surfactants*

Several recent studies have focused on the addition of lipophilic emulsifiers and their impact on the crystallization of the fat blend and the stability of the whipping cream. For example, Li et al. examined the addition of sorbitan monostearate (Span-60) with an HLB of 4.7, sucrose ester of stearate S-170 with an HLB of 1.0, lactic acid esters of monoglycerides (LACTEM) with an HLB of 4.0, sorbitan monooleate (Span-80) with an HLB 4.3, and sucrose ester of oleate O-170 with an HLB of 1.0, into a fat blend composed of AMF and HPKO in a 3:2 ratio [43]. The results showed that Span-60 and S-170, which are rich in stearic acid, exhibited strong nucleation-inducing abilities and good emulsifying properties. These emulsifiers formed tiny and uniform crystals in fat blends, resulting in small and ordered fat globules in emulsions and stable foam structures in whipped cream. LACTEM showed moderate effects due to its weaker nucleation-inducing ability. In contrast, Span-80 and O-170, which are rich in oleic acid, had poor nucleation-inducing abilities and emulsifying properties, leading to loose crystals and a decrease in whipped cream stability. The study also found that the crystallization temperature and the specific fatty acid composition of the emulsifiers significantly influenced the crystallization behavior and stability of the whipped cream. Thus, stearic acid-based emulsifiers, particularly Span-60 and S-170, are effective in enhancing the stability and sensory properties of whipped cream by promoting the formation of tiny and uniform crystals and stable foam structures. LACTEM also improves stability but to a lesser extent. Oleic acid-based emulsifiers, Span-80 and O-170 are less effective due to their poor nucleation-inducing abilities and emulsifying properties [43].

Another study used sucrose ester S370 (HLB 3, 80% sucrose diester and triester, 20% sucrose monoester, palmitic acid–stearic acid = 3:7) and the authors examined the effect on HPKO fat crystallization and whipping properties [44]. The addition of S370 had a minimal effect on droplet size, interfacial protein concentration, and the viscosity of the emulsions before whipping. However, S370, acting as a heterogeneous nucleating agent, induced nucleation, led to the formation of smaller crystals in the bulk mixture and emulsions. These smaller crystals, which could not effectively pierce the fat droplet interface, retarded fat coalescence during the whipping process. Consequently, the addition of S370 resulted in a lower degree of fat coalescence, a reduced interfacial protein concentration, a shorter optimum whipping time, and a decrease in the overrun of whipped cream. At lower S370 concentrations (<0.05 wt.%), large crystals and an inhomogeneous fat crystal network led to lower firmness and bubble stability in whipped cream samples. At 0.05 wt.% S370 concentration, a homogeneous crystal network resulted in whipped cream with high firmness and bubble stability. However, at higher S370 concentrations (>0.05 wt.%), a lower interfacial protein concentration and smaller crystals were observed, leading to a weaker aerated structure and less stable whipped cream.

The impact of different monoglycerides on the whipping abilities and partial coalescence behaviors of dairy aerated emulsions based on AMF was also studied recently focusing on three commercial monoglycerides: glyceryl monostearate (GMS), glycerol monooleate (GMO), and glycerol monolaurate (GMLa) [45]. The study found that low levels of monoglycerides (<0.25%) have minimal effects, while higher concentrations (0.50–0.75%) significantly improve the whipping performance by enhancing fat crystallization and interfacial properties. Moreover, GMS and GMLa were effective in improving the whipping abilities of aerated emulsions by enhancing fat crystallization and reducing interfacial protein concentrations. In contrast, GMO had less impact on these properties, resulting in poorer whipping performance.

Moreover, the impact of polyglycerol ester (PGE) on the crystallization of fat blends (AMF:HPKO = 3:2) and the interfacial properties, and how these factors influence the quality of whipped cream, was recently studied (Figure 2) [29]. The PGE used was a fully esterified polyglycerol, with an average glycerol polymerization degree of 3 and a fatty acid composition of only stearic acid. The results indicate that PGE significantly influences the thermal behavior and crystal morphology of fat blends. At concentrations below 0.10 wt.%, PGE coexists with sodium caseinate at the oil–water interface, reducing interfacial tension and increasing the interfacial elasticity modulus (E') and interfacial protein concentration, thereby enhancing emulsion stability. However, at concentrations above 0.10 wt.%, PGE displaces NaC, leading to a decreased E' and a decrease in interfacial protein concentration, resulting in larger, irregular fat globules and emulsion instability. At low concentrations, PGE accelerates nucleation, forming small and uniform crystals, while at high concentrations, its nucleation inducing ability reach a plateau. Whipped cream properties, including partial coalescence, interfacial protein concentration, firmness, and stability, are optimized at PGE concentrations between 0.05 wt.% and 0.10 wt.%.

By comparing these studies, we can conclude that both the nature and concentration of the lipophilic emulsifier play crucial roles. Lipophilic surfactants with unsaturated chains (low melting temperature), such as GMO, Span 80, and sucrose ester O-170, are less effective at enhancing emulsifying and whipping properties due to their limited nucleation-inducing abilities, as they form a fluid layer at the interface [46]. In contrast, surfactants with stearic chains (with high melting temperature) are highly effective at improving the stability and sensory qualities of whipped cream by promoting the formation of small crystals and therefore a stable foam. These findings offer valuable insights for producing high-quality whipped cream through the addition of a controlled surfactant. However, the reader needs to keep in mind that the exact mechanism of these long-chain LMWS is still not clearly described in the literature, which could explain some of the discrepancies. Indeed, depending on the LMWS used and the process of emulsion production, fat crystallization may nucleate at two different locations: inside the oil droplets or at the surface of the droplets [46].

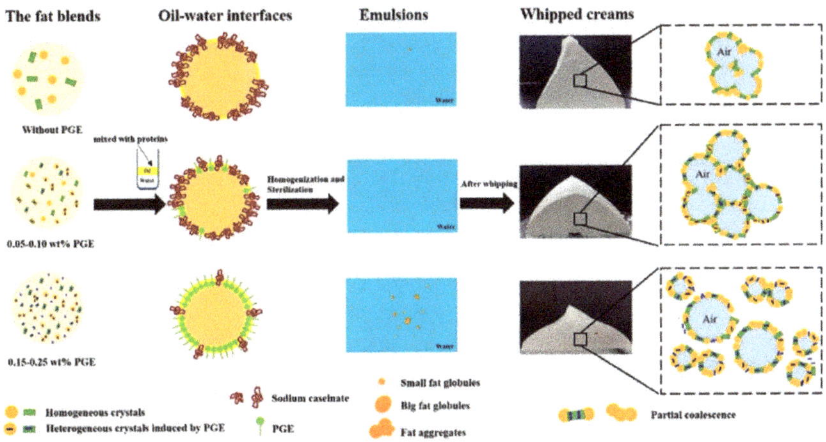

Figure 2. Schematic representation of the stabilization mechanisms of the oil fat droplets in the absence or presence of PGE at different concentrations, showing the replacement by PGE of sodium caseinate at the interface. The effect on the resulting foam after whipping is illustrated. Reproduced from [29]. Copyright 2023, Elsevier.

3.2. *Native or Modified Dairy and Plant-Based Proteins*

Proteins can fulfill diverse functions in whipping cream, acting as emulsifiers, stabilizers, viscosity-enhancers, or foaming agents. Bovine milk proteins are commonly employed in aerated dairy products to facilitate the formation of emulsions and stabilization. Caseins, which constitute approximately 80% of milk proteins, are preferred due to their conformational structure, which ensures their surface properties. For example, in the presence of LMWS, whey proteins are more prone to desorb due to their reduced flexibility (globular structure) in comparison to caseins (micellar structure) [10]. The findings of Marinova et al. indicate that casein is a more effective stabilizing agent than whey protein, forming a thicker and more resilient interfacial layer around the fat globules of WC and reducing surface tension at the oil–water interface [33]. The presence of proteins can influence several parameters of whipping creams, including the size of the fat droplets, the viscosity, the emulsion stability, and the foam stability. These effects are dependent on the concentration of proteins and their origins [10].

Whipping creams are traditionally based on dairy products. Unlike native dairy cream, recombined dairy cream (RDC) is made by combining dairy ingredients with emulsifiers and can be modified for various food applications. Bovine milk proteins, including caseins, are commonly used in aerated dairy products to facilitate the formation and stability of the emulsion. Caseins exist as micelles and can be processed into concentrate (MCC) or isolate (MCI) in their native state or modified using different technologies, including calcium caseinate (CaC) and sodium caseinate. A recent study explores how these different forms of casein affect the properties of RDC [47]. The authors investigated the impact of these three casein products (MCC, CaC, and NaC) on the stability and whipping properties of RDC, varying the concentrations of the casein products (0.5%, 1.5%, and 2.5%, *w*/*w*). The mean diameter of the MCC was found to be approximately equivalent to that of native micelles, whereas the presence of micelles was not observed in caseinates, likely due to protein disaggregation [47]. Instead, they formed large aggregates, which contributed to their relatively large particle sizes, with the CaC exhibiting the largest particle size. The highest droplet size and surface protein concentration in RDC were observed in the MCC sample, followed by CaC, and then NaC. As the protein concentration increased from 0.5% to 1.5%, the fat size decreased and the surface protein concentration increased for all samples. It was determined that an increase in casein concentration generally resulted in enhanced RDC stability attributed to the smaller particle sizes and high viscosity. However, this was not observed in the 2.5% CaC cream. Higher protein concentrations typically resulted in an increase in viscosity, with sodium caseinate creams demonstrating the highest viscosity at each concentration level. The degree of partial coalescence was found to be higher in MCC and CaC creams, which contributed to the enhancement of whipping properties. However, NaC cream was unable to be whipped into a stable aerated foam. In contrast, the MCC and CaC creams exhibited excellent whipping properties, indicating that MCC may be a more suitable ingredient for RDC preparation. The whipping properties of MCC and CaC creams were observed to be optimal at higher concentrations, whereas NaC creams were unable to form stable foams. The firmness and serum loss measurements indicated that the MCC and CaC creams at a 2.5% concentration exhibited the greatest stability against syneresis. The study's findings indicate that MCC is the most suitable casein for RDC preparation, exhibiting superior whipping properties that could probably be attributed to the larger steric repulsions [47]. Therefore, the properties of RDC can be modified and improved using different dairy ingredients.

In consideration of the market trends, consumer perceptions regarding health and safety, and commercial considerations, recent studies have focused on plant-based whipping creams, particularly in recent years. The challenge lies in balancing the stability of the emulsion and the foaming ability required for whipping cream by using plant proteins to work in the same way as dairy proteins. Soy protein, an alternative to milk proteins, has excellent nutritional value and emulsifying properties. However, its less flexible structure limits its emulsifying functionality. The enzymatic modification of soy protein isolates (SPIs) to produce hydrolysates has been explored to increase their emulsion properties. The study by Wang et al. examined the effects of various soy proteins and their hydrolysates on the stability and texture of recombined soy-based whipping cream (RSWC) [5]. The study employed a range of soy protein types, including soy protein isolate (SPI), 7S and 11S fractions, and their hydrolysates, namely soy protein hydrolyzed by pepsin (SPHPe) and soy protein hydrolyzed by papain (SPHPa) [5]. The whipping cream emulsions were prepared by combining the fat phase (20% of coconut oil) with the water phase containing the soybean oil body (28%), different soybean proteins (4%), sucrose (13%), and deionized water (35%). Soybean oil bodies are natural sources of pre-emulsified oil. The study shows that different soy proteins significantly affect the physical characteristics and whipping properties of RSWC. The lowest apparent viscosity and shortest whipping time were observed for SPHPa, indicating that this is the most readily whipped. However, it demonstrated the poorest stability, indicating that the hydrolysates produced small polypeptides that reduced viscosity. Additionally, it exhibited the highest overrun, indicating that it incorporated the greatest amount of air during whipping. Furthermore, SPHPe demonstrated favorable overrun characteristics while exhibiting enhanced stability, rendering it a more suitable option for industrial applications. SPHPe forms a robust protein network that effectively resists foam collapse and fat destabilization, which are crucial for achieving high-quality whipped cream. The confocal micrographs indicated that the small peptides derived from hydrolyzed proteins displaced oil body proteins at the interface, resulting in the partial destabilization of fat globules and enhanced fat coalescence [5]. Hydrolyzed soy proteins, particularly SPHPe, can be effectively used in the production of an RSWC with sufficient overrun and whipping stability, offering a possible alternative to traditional dairy-based whipping creams.

To summarize, all the studies revealed that to enhance the foaming characteristics of whipped creams, it is optimal to utilize milk proteins, particularly caseins, in their native state. Conversely, when plant-based proteins are employed in their original form, due to their limited solubility, the resulting properties are observed to be inferior. In order to enhance the functionality of plant-based proteins, it is necessary to subject them to a series of modifications, including hydrolysis.

3.3. *Surface-Active Agent Complexation*

The use of only surface-active agents, such as plant-based proteins or LMWS, is typically insufficient to stabilize both WC emulsion and the subsequent foam formed, as illustrated by the aforementioned work. The incorporation of a mixture of emulsifiers or the formation of a complex such as gel-like particles or Pickering emulsion represents a viable approach to enhance the functionalities of surface-active agents in WC. A Pickering emulsion is a type of emulsion stabilized by solid particles, which adsorb at the interface of immiscible liquids (e.g., oil and water) to prevent coalescence. These particles create a physical barrier, enhancing emulsion stability.

A recent study investigated the interaction and synergetic effects of SPI and its hydrolysates with varying concentrations of monoglycerides at the air–water/oil interfaces in recombined low-fat whipping cream. The WC was formulated with 20% palm oil, 18% carbohydrate, 0.22% stabilizers, and 0.25–1.00% monoglycerides. The proteins used included native soy protein isolate (NSPI), commercial soy protein isolate (CSPI), SPHPe, SPHPa, and NaC. The results indicated that increasing the monoglyceride concentration improved the texture, whipping properties, and stability of the whipped cream by displacing the adsorbed protein from fat globules, building a firmer structure of fat aggregates, and stabilizing trapped air bubbles. SPHPa whipped cream showed a similar overrun, stability, and texture to NaC, while SPHPe led to low overrun and weakened structure due to the high proportion of β-conglycinin. SPHPa whipped cream showed similar properties to NaC for the studied formulations. The high proportion of β-subunits in SPHPe reduced the partial coalescence, resulting in a low overrun and weak structure. The study provides insights into the interaction between soy protein molecules and monoglycerides at the oil–water interface and their impact on the properties of low-fat whipped cream [5]. The results contradict the study of Fu et al. described previously [48]. We can suppose that the differences could be due to the difference in the oil phase used (coconut oil versus palm oil), the presence of monoglycerides, and/or the presence of a soybean oil body in the water phase. More work needs to be carried out on soy protein isolates and hydrolysates to better understand the differences between the studies.

Another recent study examined the foaming properties of whipped cream stabilized by faba bean protein isolate microgel particles (FPIMPs) as a substitution for sodium caseinate (Figure 3) [49]. The faba bean protein was employed as a microgel particle to guarantee its functionality, as it has poor solubility in its native state due to its compact structure, which is a common trait among plant proteins [50]. This lack of solubility among plant proteins can negatively affect their emulsifying and foaming properties. The utilization of microgel particles presents a novel avenue for the exploitation of plant proteins in contexts in which complete protein dissolution is a prerequisite [51]. In addition to faba bean, other proteins, including pea, peanut, and soybean, were employed in the fabrication of microgels for the stabilization of emulsions, with notable success [48,52–54]. Microgel particles are soft solid particles that are formed via the mechanical disruption of gelled polymers through, for example, homogenization. Proteins, as a natural polymer, can be transformed into gels through a thermal cross-linking, enzymatic cross-linking, or a combination of both, and subsequently transformed into protein microgel particles. The recipe for cream emulsions consisted of 23% palm kernel stearin, 10% sugar, 3% glucose syrup, 0.1% potassium dihydrogen phosphate, 0.6% emulsifiers, and 0.1% xanthan. Additionally, the following ingredients were utilized: (0.08%) guar gum, (0.05%) sodium carboxymethyl cellulose, and FPIMPs with varying protein concentrations (0%, 0.2%, 0.5%, 0.8%, 1.1%, or 1.4%). A control sample was produced with the same recipe in which the FPIMPs were substituted with 0.8% sodium caseinate. The physical stability of the emulsions was initially elucidated. The measurement of fat globule distribution demonstrated a correlation between the average fat size and the FPIMPs content, with an observable increase in size as the FPIMPs concentration increased (Figure 3). This phenomenon can be attributed to the propensity of Pickering particles to form aggregates with a high concentration of FPIMPs, which is detrimental to emulsifying stability. The lowest fat globule size was obtained with NaC compared to other creams with FPIMPs. These results, as reported by other researchers, indicate that sodium caseinate is more effective than protein particles in preventing droplet flocculation in whipping cream [26]. It can be observed that an increase in FPIMPs results in an elevated viscosity of whipping creams. This phenomenon can be ascribed to the enlargement of fat globules [43]. However, sodium caseinate was more easily displaced from the oil–water interface than protein particles, which have a high absorption energy [55]. The measurement of foaming properties demonstrated that the creams containing NaC exhibited the most optimal foamability results. The replacement of proteins at the interface by LMWEs modified the overrun rate and whipping time. In the case of creams containing FPIMPs, the highest overrun and lowest whipping time were observed with a high FPIMPs content (0.8–1.4%) (Figure 3). With an increase in FPIMPs content, the number of protein particles in the liquid phase increases, playing a beneficial role in foam stabilization. This is achieved via the formation of a network that prevents drainage and ensures the rigidity of the foam structure [23].

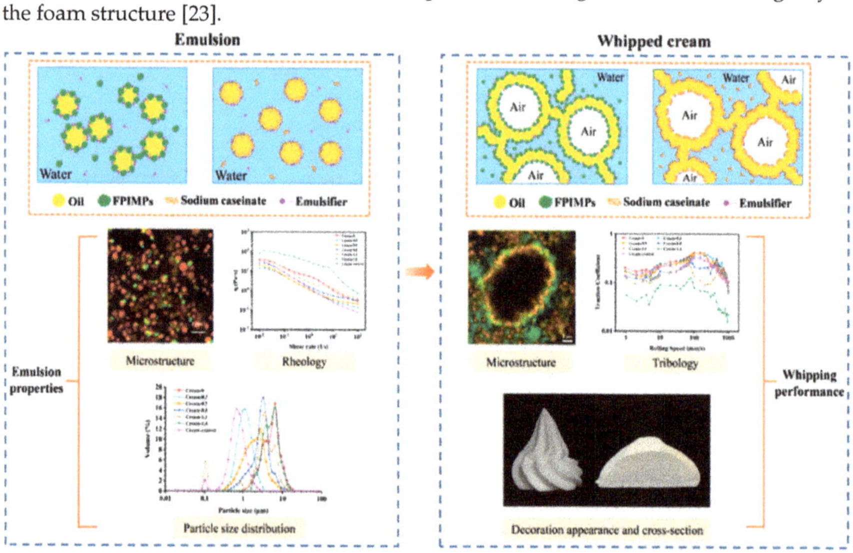

Figure 3. Schematic representation of the stabilization mechanisms of the oil fat droplets with sodium caseinate versus faba bean microgels and their corresponding whipping properties. Reproduced from [49]. Copyright 2024, Elsevier.

3.4. Pickering Emulsion Formation

Another recent trend is to use particles or complexes to form Pickering emulsions, which are known for their stability due to the irreversible adsorption of particles on the interface. Zein nanoparticles (ZNP) have garnered attention due to their hydrophobic nature. Cao et al. investigated the potential of zein colloidal particles (ZCPs) to enhance the mechanical properties of whipped cream made with highly polyunsaturated oils

(HPUO) [26]. The researchers found that the addition of ZCPs significantly improved the overrun and foam stability of the whipped cream during the whipping process. The ZCPs-reinforced whipped cream exhibited desirable hardness and maintained its shape better than the control samples without ZCPs. Unlike conventional whipped cream, which relies on partial coalescence of fat globules for hardness, the ZCPs-containing samples did not show partial coalescence during whipping. Instead, the particle-packed droplets in the emulsion tended to aggregate, enhancing the structure of the triphasic air–HPUO–water system. Thus, the ZCPs reinforced the whipped cream. The authors suggested that the ZCPs stabilize oil droplets by forming a rigid armor around them, which prevents partial coalescence and promotes the formation of a gel-like network. This network, along with the competitive adsorption of emulsifiers and ZCPs at the oil–water and air–water interfaces, stabilizes the air bubbles and enhances the foam's rigidity. The aggregation of ZCPs-packed droplets further reinforces the network, contributing to the improved mechanical properties of the whipped cream. This study demonstrated that ZCPs could effectively improve the mechanical properties of HPUO-based whipped cream, making it a potential strategy for developing healthier aerated foods with non-hydrogenated and less-saturated oils.

Previous studies have demonstrated the successful stabilization of emulsions using zein-based nanoparticles combined with other substances like polysaccharides [56] and phenolic acids [57]. Grossi et al. developed a plant-based whipped cream by utilizing plant proteins and unsaturated oils formulated through a zein nanoparticles–sodium stearate complex (Figure 4). The complexes were combined with monoacylglyceride (MAG)-based oleogels to increase the solid fat content of vegetable oils [25]. In this study, the researchers functionalized insoluble zein with sodium stearate, significantly enhancing its solubility, emulsifying, and foaming properties. However, the use of a high amount of sodium stearate can increase potential consumer concerns. The study aimed to reduce the amount of sodium stearate by forming zein–sodium stearate nanoparticles and using oleogels as the oil phase. The hypothesis was that these nanoparticles could stabilize Pickering emulsions and incorporate air upon whipping, with the gas bubbles being stabilized by partial coalescence of the fat droplets, resulting in a stable plant-based whipped cream. A form of ZNP-SS was produced and characterized in terms of molecular interactions, wettability, and surface activity. Moreover, their ability to stabilize oil-in-water emulsions and foams in the presence or absence of MAG was evaluated. The results show that the ZNP-SS complex was stabilized through electrostatic and hydrophobic interactions. An increase in the amount of sodium stearate caused an increase in its hydrophilicity, which improved the interfacial properties of ZNP-SS. Corn oil-in-water emulsions were successfully stabilized by all the ZNP-SS, while only the ones with the highest amount of sodium stearate (ZNP-0.6) were able to create stable whipped creams due to the partially coalesced oil droplets that occurred in the presence of MAG. Thus, this study shows an innovative approach to developing stable plant-based whipped cream, representing a significant step towards creating low-saturated-fat products.

Another study investigated the synergistic effects of xanthan gum and β-cyclodextrin (β-CD) on the properties and stability of vegetable oil-based whipped cream stabilized by kidney bean protein aggregates (Figure 5) [58]. Xanthan gum is commonly used as a hydrocolloid stabilizer in O/W emulsions and whipped cream, contributing to long-term storage stability. It is often combined with proteins or other polysaccharides to improve emulsion stability. Xanthan gum, carrying negative charges, can interact electrostatically with positively charged macromolecules in whipped cream, enhancing viscosity and foam stability. β-Cyclodextrins, representative of oligosaccharides, have attracted attention due to their ability to enhance emulsion stability due to their amphiphilicity [59]. However, their low molecular weight limits their shaping properties in whipped cream, necessitating the auxiliary effect of large molecular polysaccharide components. The interaction between oligosaccharides and polysaccharides in composite systems, which can regulate colloid properties and improve Pickering emulsion stability, remains underexplored, particularly in vegetable oil-based whipped cream systems. The study examines the effect of β-cyclodextrin's unique cone-like microstructure on maintaining the performance and stability of vegetable oil-based whipped cream by acting as a bridge between the hydrophilic and hydrophobic components. The materials used in the study include xanthan gum, β-cyclodextrin, kidney beans, monoglycerides and diglycerides of fatty acids, sodium stearyl lactate, corn syrup, soybean oil, and sucrose. The preparation of kidney bean protein aggregates (KPAgs) involves alkaline extraction, isoelectric precipitation, freeze-drying, and various treatments to obtain the desired protein solution. The preparation of xanthan gum and β-cyclodextrin complexes, emulsions, and whipped creams involves dissolving XG and β-CD in distilled water at various ratios, combining them with KPAgs solution, corn syrup, and sucrose, and creating a Pickering emulsion system. The emulsion is then whipped to form whipped cream (Figure 5). The study demonstrates that an appropriate XG to β-CD ratio significantly enhances the hydrogen bonding, resulting in a dense three-dimensional network structure with small, evenly distributed emulsion droplets. The unique tapered microstructure of β-CD acts as a bridge between the hydrophilic and hydrophobic components, preventing the aggregation of oil droplets and establishing a flexible support system, while XG supports the Pickering emulsion system (Figure 5). This study provides valuable insights into the production of entirely plant-based whipped cream by texturing highly unsaturated oils, addressing the issue of the inadequate intake of unsaturated oil for individuals consuming excessive animal-derived fats.

A very interesting study explored the potential to use lactic acid bacteria (LAB) as structural building blocks in non-fat whipping cream analogs [60]. The research focused on replacing saturated fat content in whipping cream with LAB combined with hydroxypropyl methylcellulose and sodium caseinate salt. Microorganisms, including food-grade yeasts and non-pathogenic LAB, were reported to stabilize emulsions, double emulsions, and dry foam via a Pickering effect. The structural role of microorganisms as Pickering particles is greatly dependent on their surface chemical compositions and physicochemical properties, such as cell hydrophobicity, surface charge density, and aggregation properties. Thus, the aim of this work was to determine whether the structural functionality of solid fat globules, achieved through Pickering effects and partial coalescence, could be replicated using edible LAB with selected surface and aggregation properties. Two strains of LAB, *Lactobacillus delbrueckii subs. lactis* ATCC 4797 (LBD) and *Lactobacillus crispatus* DSM20584 (LBC), with different surface properties, were combined with NaC and HPMC to create a series of whipping cream-like suspensions. The study found that the overrun of the whipped cream was mainly attributed to the use of HPMC and NaC rather than the bacteria due to the faster adsorption of molecular surfactants compared to micron-sized bacterial cells. However, the drainage stability was systematically dependent on the strain that was used, with LBD samples allowing for nearly twice the drainage of the LBC samples. The viscoelasticity of the whipped cream was measured, showing that all samples exhibited higher storage modulus (G') values than loss modulus (G'') values, indicating a more solid than liquid character. Formulations with LBC showed much higher G' values than LBD samples, suggesting stronger bacterial aggregation and more solid-like properties. The aggregation of LBC was primarily enhanced by the NaC component, while for LBD samples, HPMC was more capable of promoting bacterial aggregation (Figure 6). The microstructure of the whipped cream was investigated using confocal laser scanning microscopy, revealing that the hydrophobic LBC exhibited almost full coverage at the air–water interface, indicating a stronger Pickering capability than the hydrophilic LBD. The viability of bacteria in the whipped cream was quantified, showing that neither the added components nor the whipping process had detrimental effects on bacterial viability. In conclusion, non-fat whipping cream analogs with a high overrun and drainage stability were formulated using LAB as natural structural building blocks combined with NaC and HPMC components. The more hydrophobic LBC strain created whipped cream with a higher stiffness and liquid retention ability than the hydrophilic LBD strain due to the strong Pickering stabilizing capacity and network-forming ability of LBC bacteria. The aggregation of the two strains was enhanced to differing extents by the added components, with NaC better promoting the aggregation of LBC and HPMC inducing a stronger aggregation of LBD (Figure 6). The study opens the possibility of using natural LAB as both structural building blocks and probiotic food components.

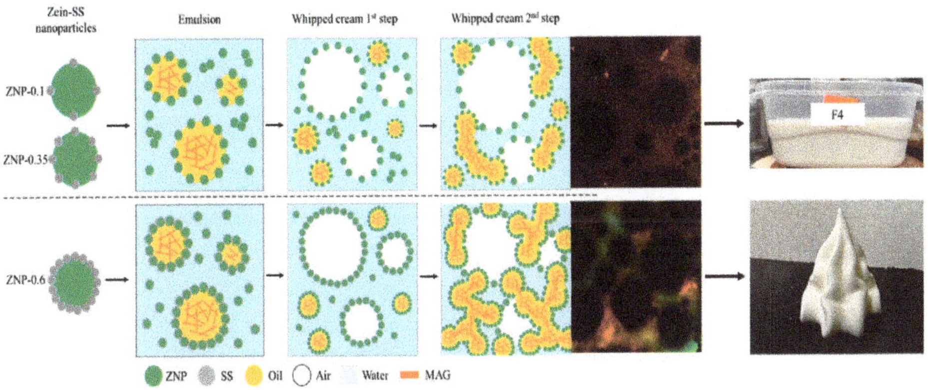

Figure 4. Schematic representation of the stabilization mechanisms of the oil fat droplets containing an oleogel based on monoglyceride using a complex based on zein nanoparticles (ZNP) and sodium stearate (SS). The effect of the whipping process on the resulting foam stabilization and properties is illustrated as a function of the quantity of sodium stearate used in the complex with ZNP. Reproduced from [25]. Copyright 2024, Elsevier.

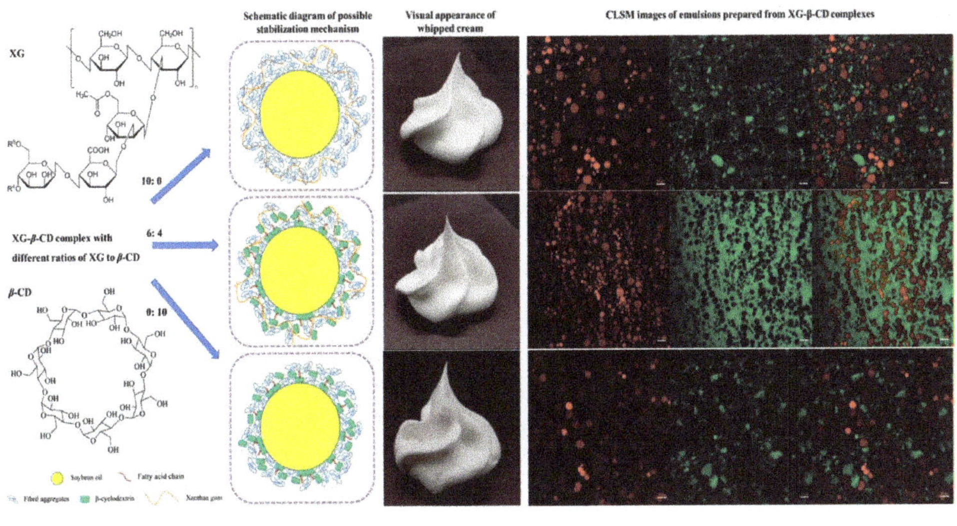

Figure 5. Schematic representation of the stabilization mechanisms of the oil fat droplets using a complex based on XG and β-CD depending on the ratio of the two components, the corresponding confocal laser scanning microscopy images, and photographs of the resulting whipping cream. As the concentration of β-CD in the complex increased, the droplet size of the emulsion system initially decreased and then increased, as confirmed by the average particle size measurements. At a XG–β-CD ratio of 6:4, the emulsion droplets were spherical, small in size, and uniformly distributed without any coalescence. At this ratio, red fluorescence was observed, encircling the surface of the oil droplets, forming a fluorescent ring. This indicated that the majority of XG–β-CD complex particles were tightly bound to the surface of the oil droplets, creating a spatial barrier that prevented agglomeration in the Pickering emulsions. Reproduced from [58]. Copyright 2024, Elsevier.

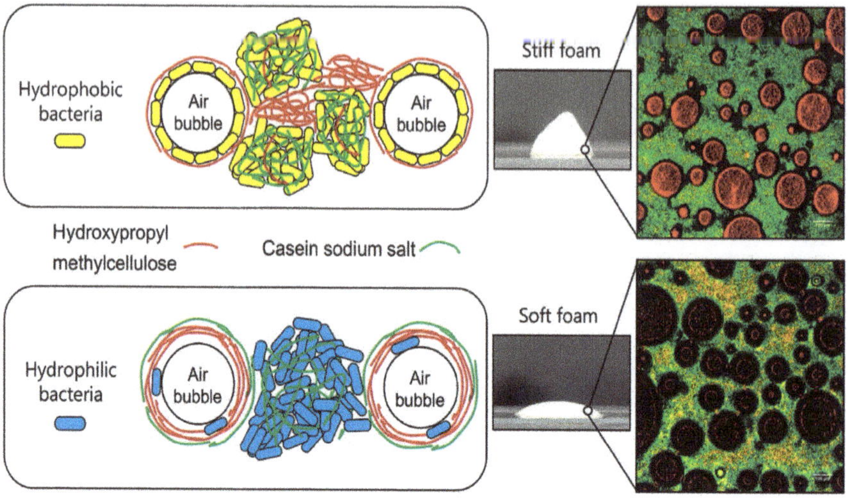

Figure 6. Schematic representation of the stabilization mechanisms of the air bubbles via the presence of hydrophobic or hydrophilic bacteria in the presence of hydroxypropylmethycellulose and casein sodium salt and the corresponding confocal laser scanning microscopy images of the resulting foams after whipping, as illustrated by the photographs. Reproduced from [60]. Copyright 2023, Elsevier.

4. Effect of Stabilizers in the Continuous Phase

Reducing the fat content in whipped cream often results in weaker textural properties, such as reductions in firmness, viscosity, overrun, and stability against drainage. To address these issues, the addition of different stabilizers has been explored, such as fibers and hydrocolloids, to improve the texture of low-fat whipped cream. Various fat replacers, including modified whey protein isolate, rice starch, different seed gums, and glucose, have also been used to enhance the textural properties of low-fat whipped cream. Not only does low-fat whipped cream need to be achieved, but also the low sugar content WC what increase the complexity of the formulation. Therefore, a current research focus is developing whipped cream analogs with low-fat content or exploring new additives with the desired properties. The stabilization of whipped cream is a critical technical problem that has attracted much attention. Exploring the interactions between food components and finding new food additives prepared from food-grade materials are promising solutions to this problem in the food industry.

4.1. Effect of Sugar Addition

The study by Zeng et al. investigates the impact of glucose and corn syrup on the physical characteristics and whipping properties of vegetable-fat-based whipped creams [4]. This study analyzes how glucose and corn syrup affect the physical and foaming properties of vegetable-fat-based whipped creams. The research focuses on glucose and corn syrup and how varying concentrations of these sugars affect the emulsions' interfacial protein concentration and apparent viscosity, as well as the overall stability and firmness of the whipped creams. It was observed that corn syrup, which contains maltodextrin, resulted in an increase in interfacial protein concentration and a reduction the coalescence of oil droplets compared to glucose. Maltodextrin protects the absorbed proteins during freezing and slows the formation of a continuous network during whipping, resulting in reduced fat coalescence and a lower firmness in the whipped cream. Despite these differences, whipped creams with corn syrup still exhibit relatively good stability, especially at higher sugar concentrations. The addition of sugars, particularly glucose and corn syrup, influences the whipping properties of the cream. Higher sugar concentrations lead to shorter optimum whipping times, an increase in fat coalescence droplets, and increased firmness and stability in the whipped creams. These effects are attributed to the sugars' ability to increase the interfacial protein concentration and the apparent viscosity of the emulsions. In terms of practical applications, the study suggests that the partial replacement of glucose with corn syrup or maltodextrin could improve the stability and foaming properties of WC. This finding is significant for the development of low sugar whipped cream products, which are increasingly in demand due to health concerns related to high sugar intake.

4.2. *Effect of Polysaccharides Addition*

A recent study demonstrated how cellulose extracted from sugarcane bagasse using chemicals before undergoing a specific electrohydrodynamic (EHD) treatment could be a good stabilizer agent for whipping cream [61]. Cellulose, a biopolymer composed of repeating β-d-glucopyranose units linked via 1→4 glycosidic linkage, is the most abundant organic polymer and an inexpensive source of food additives [62]. Due to its highly crystalline nature, cellulose is often modified to improve its techno-functional properties such as its water absorption capacity, viscosity, and gelling ability [63]. Modified cellulose has been used as a fat replacer in various dairy products, including dairy beverages, yogurt, ice cream, and whipped cream. However, the literature on the application of cellulose modified using electrical field treatment in food products is limited. EHD treatment is a high-voltage, non-thermal food processing method that generates corona discharge, creating an electric wind that induces mass transfer and microstructural changes in biological materials. EHD treatment has been used for food-drying, nutrient encapsulation, and the extraction of food compounds [64]. The cellulose extracted from sugarcane bagasse and treated with EHD showed a partial transformation from a microcrystalline to amorphous state, enhancing its water absorption capacity and viscosity. The authors compared the addition of modified and non-modified cellulose to whipped cream with varying fat concentrations. The results showed that an increase in cellulose and fat concentrations improved the viscosity, firmness, overrun, and stability against syneresis of the whipped cream. Whipped cream samples containing modified cellulose exhibited superior textural and color qualities compared to those with non-modified cellulose. The sample with 1.5% modified cellulose and 30% fat showed preferable stability, texture, and color properties. Thus, EHD-treated cellulose fibers enhanced the stability, texture, and color of whipped cream, making it a viable option for the production of lower-fat dairy products.

A recent research article investigated the impact of basil seed gum (BSG), a novel polysaccharide, known for its thickening and emulsifying properties [65]. The authors compared the effects of BSG and κ-carrageenan on the rheological, textural, and structural properties of whipping cream [65]. The results indicate that whipping cream containing BSG exhibits a higher viscosity compared to whipping cream with κ-carrageenan. BSG significantly improved the cream's viscosity, demonstrating a strong capacity to enhance the rigidity of the emulsion. A rheological analysis showed that BSG improves the low-frequency dependence of the elastic modulus, indicating a more rigid structure. The study also found that BSG can form a pseudo-gel network by adsorbing protein segments at the oil–water interface, creating a stronger molecular network in the whipping cream. In contrast, whipping cream with κ-carrageenan showed some degree of flocculation, likely due to non-adsorbed polysaccharides or proteins. A microstructural analysis using scanning electron microscopy revealed that whipping cream with BSG contained more voids and pores, which decrease with an increase in fat content, enhancing the foam structure. The study concludes that synergistic interactions between proteins and polysaccharides, such as BSG and κ-carrageenan, promote the development of a cross-linked network, improving the emulsifying properties of WC. However, further research is needed to explore the mechanisms behind BSG's interactions at the oil–water interface and its impact on the microstructure of whipped cream.

A first study investigating the potential functions of hydroxypropyl methylcellulose (HPMC) in whipping cream was conducted, finding that HPMC displaced proteins from the interface and improved the textural and foaming properties of whipped cream [66].

Then, a second study investigated the emulsifying and foaming properties of mixing polysaccharide dispersions, specifically focusing on the effect of the insoluble soybean fiber (ISF) and HPMC ratio [67]. ISF is the main component of okara, obtained in large quantities during the production of soy-based products, and might provide great valorization opportunities for both the application of agricultural byproducts and the exploration of novel food additives [67]. The study explored how the interactions between ISF and HPMC influence the processing and characteristics of food systems such as whipped creams. In this study, HPMC was partially replaced by ISF, and O/W emulsions and whipped cream were prepared. The mixture of ISF/HPMC exhibited a good foaming performance after whipping, and the foaming performance was enhanced as the proportion of ISF increased. In the O/W emulsion system, the overall stability, including the acid and salt resistance, was maintained, while the heat stability and centrifugal stability of the O/W emulsions first increased and then decreased as the proportion of ISF increased. This indicated that the partial replacement of HPMC with ISF could help to improve the emulsion stability. In the whipping cream system with an increase in the proportion of ISF, the overrun, foam stability, and cream stability first increased and then decreased. The size of the fat bubbles decreased, and the elastic modulus first increased and then decreased. This indicated that the addition of an appropriate amount of ISF strengthened the network structure of whipping cream and enhanced the stability of foam and cream, thus showing its excellent application potential in whipping cream. Thus, ISF could be used to partially replace HPMC to prepare O/W emulsions and whipped cream.

As described above, hydrocolloids are efficient ingredients to improve the stability of whipping creams, but little information is available on their impact on the nutritional properties and lipid digestion. A recent study focused on the use of carrageenan and pectin in whipping cream and explored their potential to not only improve the textural properties of whipped cream, but also to retard lipid digestion [22]. Indeed, hydrocolloids can induce the flocculation of lipid droplets in the gastrointestinal tract, thereby decreasing the lipid digestion rate. The hydrocolloid solutions of carrageenan and pectin were prepared by dispersing the hydrocolloids in distilled water, followed by stirring and refrigerating for complete hydration. Fresh cream with a fat content of approximately 40% was adjusted to a 30% fat content by adding raw skimmed milk. The hydrocolloid solutions were then added to the cream, which was pasteurized, homogenized, and refrigerated to promote fat crystallization. The whipping cream was then aerated using a stand mixer, and the samples were examined for their in vitro lipid digestion properties [22]. The study found that the addition of hydrocolloids reduced the average volume–surface diameter of the lipid droplets and increased the net negative charges, indicating improved homogenization efficiency. Both pectin and carrageenan enhanced the foaming properties of the whipped cream, with pectin generally exhibiting higher overrun values. The addition of hydrocolloids also significantly reduced serum loss, indicating an improvement in foam stability. In terms of lipid digestion, both hydrocolloids caused the flocculation and/or coalescence of droplets when exposed to pH changes during the digestion model, resulting in creaming and phase separation of the emulsions. Carrageenan was found to be more effective in reducing the rate of lipid digestion, as evidenced by the reduced amount of released free fatty acids compared to pectin. In conclusion, the study demonstrates that hydrocolloids like pectin and carrageenan can be used in whipping creams to both improve foaming properties and retard lipid digestion. These findings suggest that the rate and extent of lipid digestion can be controlled by adding certain types of polysaccharides, which may be useful for developing healthier whipped cream products with better functional properties.

4.3. *Effect of Protein–Polysaccharide Complex*

In the literature, several examples have been provided of the use of polysaccharides in combination with other food ingredients to improve the properties of whipping creams. For example, protein–polysaccharide complexes, due to their high emulsifying ability and suitable surface activity, have been considered as potential fat-replacers in low-fat whipping cream [68]. These complexes can improve the stability of oil–water emulsions by forming new additives with various interfacial properties. The combination of proteins and polysaccharides yields an emulsion with enhanced functional properties, such as improved stabilization and emulsification. For example, the study by Rezvani et al. demonstrated the effects of a protein–polysaccharide complex (sodium caseinate, carboxymethyl cellulose, and locust bean gum) on the physical and textural characteristics of low-fat whipping cream [68]. The authors used the response surface methodology to optimize the proportions of protein and polysaccharides to find the best qualitative properties including overrun, apparent viscosity, firmness, average particle size, drainage, and color parameters. Their results showed that the proposed combination of stabilizers improved the physical and sensory characteristics of low-fat whipped cream, enhancing its apparent viscosity, firmness, average particle size, and overrun while reducing drainage. A sensory evaluation revealed that the most desirable treatment optimization for low-fat cream contained 0.35% sodium caseinate, 0.15% carboxymethyl cellulose, and 0.15% locust bean gum.

Another study focused on the effect of a protein–k-carrageenan mixture [24]. A low-fat whipping cream was formulated by substituting the removed fat content with a blend containing water, SPI, and k-carrageenan gum. SPI are excellent emulsifiers that can prevent oil droplets from coalescing, while k-carrageenan can increase viscosity and reduce serum phase leakage, enhancing WC stability. The influence of SPI and k-carrageenan content on the stability of the WC was examined using a simplex–centroid design to evaluate the optimum formulation. Various properties, such as foaming, viscosity, textural properties, and emulsion stability, were determined. The physical stability of the whipped cream was evaluated by measuring overrun, viscosity, adhesiveness, and firmness. The results showed that water proportion significantly affected foamability, with a higher water content leading to an increase in foam capacity. SPI also increased foam volume, while high levels of k-carrageenan decreased overrun and increased viscosity. The stability of the foam was ensured by the presence of SPI and k-carrageenan, which provided rigidity and visco-elasticity to the films, stabilizing air bubbles. The viscosity of the samples varied, with higher values being observed in formulations with more k-carrageenan. A texture analysis showed that a higher k-carrageenan content resulted in a harder and more stable foam. Emulsion stability was maintained by the presence of emulsifiers, k-carrageenan, and SPI, which prevented phase separation. A microscopic examination revealed differences in the microstructure of the cream samples, with monodispersed emulsions showing higher viscosity. A sensory evaluation indicated that the addition of SPI and k-carrageenan improved the sensory attributes of low-fat whipping creams. The optimum fat substitute blend contained 44.5% water, 46.4% SPI, and 9% k-CG, resulting in a product with 31.5% fewer calories than full-fat WC. These two studies highlighted the possible use of a protein–polysaccharide complex for the development of low-fat, stable, and acceptable whipping cream products.

5. Conclusions and Future Directions

Whipping cream is a widely used product in the food industry, and its market has grown substantially in recent years due to increasing consumer demand for healthier and higher-quality options. This growing interest has driven substantial scientific research and innovation. In this review, we highlight the most recent literature (from the past three years) focusing on the effects of proteins, low-molecular-weight surfactants, and stabilizers in whipping cream formulations. In the future, a major focus will be on creating healthier alternatives, such as whipping creams fortified with probiotics and formulations with reduced levels of saturated fats and sugars. The development of low-fat, low-sugar, and functional whipping creams is a key priority and requires further attention. Consumers also continue to seek more "natural" foods and "clean label" products without synthetic additives, meaning that even if functional and nutritional improvements are achieved through additives, the health impact and perceived naturalness of these additives must be carefully considered. Another important trend is the growing development of plant-based substitutes for dairy creams. However, more research is needed to match the sensory and functional properties of traditional dairy creams. Ongoing studies into plant proteins (isolates, solid particles, microgels, etc.) are essential to create alternatives that replicate the texture, stability, and overall quality of dairy-based whipped creams. All the studies confirmed that to enhance the functionality of plant-based proteins, it is necessary to subject them to a series of modifications, including hydrolysis, particle formation, or microgels, to create a gel-like network or to use them in synergy with other emulsifiers. Furthermore, a better understanding of the role of LMWS is crucial for producing high-quality whipping cream through controlled surfactant use, as their exact mechanisms are still not fully described in the current literature. Clarifying the nucleation process—whether it begins inside the oil droplets or on their surface—will also be key to controlling the final properties of whipped creams.

In summary, the development of whipping creams with improved nutritional and functional properties, as well as non-dairy alternatives, presents both significant opportunities and challenges for future research.

Author Contributions: K.F.D.: writing—original draft, writing—review and editing. A.-L.F.: writing—review and editing, supervision. R.K. and C.C.: resources, review and editing, supervision. All authors have read and agreed to the published version of the manuscript.

Funding: This research received no external funding.

Institutional Review Board Statement: Not applicable.

Informed Consent Statement: Not applicable.

Data Availability Statement: Not applicable.

Conflicts of Interest: The authors declare no conflicts of interest.

References

1. Andrade, J.; Rousseau, D. Whipping Properties of Recombined, Additive-Free Creams. *J. Dairy Sci.* **2021**, *104*, 6487–649 [CrossRef] [PubMed]
2. Whipping Cream Market Forecast, Share, Strategies, Scope, Overview. Whipping Cream Market Size, Share and Forecas 2025–2035. Available online: https://www.marketsizeandtrends.com/download-sample/673662/?utm_source=Pulse-2-5visite (accessed on 26 September 2024).
3. Gafour, W.; Aly, E. Organoleptic, textural and whipping properties of whipped cream with different stabilizer blends. *Acta Sc Pol. Technol. Aliment.* **2020**, *19*, 425–433. [CrossRef]
4. Zeng, Y.; Zeng, D.; Liu, T.; Cai, Y.; Li, Y.; Zhao, M.; Zhao, Q. Effects of Glucose and Corn Syrup on the Physical Characteristics an Whipping Properties of Vegetable-Fat Based Whipped Creams. *Foods* **2022**, *11*, 1195. [CrossRef]
5. Wang, Z.; Liang, G.; Chen, W.; Qie, X.; Fu, L.; Li, X.; He, Z.; Zeng, M.; Goff, H.D.; Chen, J. Effects of Soy Proteins and Hydrolysate on Fat Globule Coalescence and Whipping Properties of Recombined Low-Fat Whipped Cream. *Food Biophys.* **2022**, *17*, 324–33 [CrossRef]
6. Han, J.; Zhou, X.; Cao, J.; Wang, Y.; Sun, B.; Li, Y.; Zhang, L. Microstructural Evolution of Whipped Cream in Whipping Proce Observed by Confocal Laser Scanning Microscopy. *Int. J. Food Prop.* **2018**, *21*, 593–605. [CrossRef]
7. Wang, Y.; Yuan, D.; Li, Y.; Li, M.; Wang, Y.; Li, Y.; Zhang, L. Thermodynamic and Whipping Properties of Milk Fat in Whippe Cream: A Study Based on DSC and TD-NMR. *Int. Dairy J.* **2019**, *97*, 149–157. [CrossRef]
8. Siva Kumar, S.; Balasubramanyam, B.V.; Jayaraj Rao, K.; Heartwin Amala Dhas, P.; Surendra Nath, B. Effect of Flaxseed Oil an Flour on Sensory, Physicochemical and Fatty Acid Profile of the Fruit Yoghurt. *J. Food Sci. Technol.* **2017**, *54*, 368–378. [CrossRe [PubMed]
9. Shim, S.Y.; Ahn, J.; Kwak, H.S. Functional Properties of Cholesterol-Removed Whipping Cream Treated by β-Cyclodextrin. *Dairy Sci.* **2003**, *86*, 2767–2772. [CrossRef]
10. Anihouvi, P.; Danthine, S.; Karamoko, G.; Blecker, C. *Les Crèmes Végétales: Une Alternative Aux Crèmes Laitières (Synthè Bibliographique)*; BASE: Tokyo, Japan, 2012; Volume 16.
11. Carr, N.O.; Hogg, W.F. A Manufacturer's Perspective on Selected Palm-Based Products. *Asia Pac. J. Clin. Nutr.* **2005**, *14*, 381–38 [PubMed]
12. Riaublanc, A.; Anton, M.; Mariette, F.; Georges, C.; Gravier, E.; Drelon, N.; Omari, A.; Leal-Calderon, F. Impact Des Cristaux c Matière Grasse Sur l'aptitude Au Foisonnement et La Stabilité Des Émulsions Laitières Foisonnées. *Sci. Aliment. = Food Sci. Int. Food Sci. Technol.* **2005**, *25*, 427–441. [CrossRef]

13. McClements, D.J.; Jafari, S.M. Improving Emulsion Formation, Stability and Performance Using Mixed Emulsifiers: A Review. *Adv. Colloid Interface Sci.* **2018**, *251*, 55–79. [CrossRef] [PubMed]
14. Chen, Y.; Cai, Y.; Di, Z.; Zhao, M.; Zhao, Q. Improving Quality Characteristics of Whipped Cream Based on Novel Additives: A Review of Current Status, Challenges, and Strategies. *Food Bioprocess Tech.* **2024**. [CrossRef]
15. Goff, H.D. Instability and Partial Coalescence in Whippable Dairy Emulsions. *J. Dairy Sci.* **1997**, *80*, 2620–2630. [CrossRef]
16. Moens, K.; Tavernier, I.; Dewettinck, K. Crystallization Behavior of Emulsified Fats Influences Shear-Induced Partial Coalescence. *Food Res. Int.* **2018**, *113*, 362–370. [CrossRef] [PubMed]
17. Munk, M.B.; Andersen, M.L. Partial Coalescence in Emulsions: The Impact of Solid Fat Content and Fatty Acid Composition. *Eur. J. Lipid Sci. Technol.* **2015**, *117*, 1627–1635. [CrossRef]
18. Fuller, G.T.; Considine, T.; Golding, M.; Matia-Merino, L.; MacGibbon, A. Aggregation Behavior of Partially Crystalline Oil-in-Water Emulsions: Part II—Effect of Solid Fat Content and Interfacial Film Composition on Quiescent and Shear Stability. *Food Hydrocoll.* **2015**, *51*, 23–32. [CrossRef]
19. Ihara, K.; Habara, K.; Ozaki, Y.; Nakamura, K.; Ochi, H.; Saito, H.; Asaoka, H.; Uozumi, M.; Ichihashi, N.; Iwatsuki, K. Influence of Whipping Temperature on the Whipping Properties and Rheological Characteristics of Whipped Cream. *J. Dairy. Sci.* **2010**, *93*, 2887–2895. [CrossRef]
20. Liu, N.; Cui, J.; Li, G.; Li, D.; Chang, D.; Li, C.; Chen, X. The Application of High Purity Diacylglycerol Oil in Whipped Cream: Effect on the Emulsion Properties and Whipping Characteristics. *CYTA-J. Food* **2019**, *17*, 60–68. [CrossRef]
21. Scopus-Document Details-Effects of Sodium Caseinate and Whey Proteins on Whipping Properties and Texture Characteristics of Whipped Cream. Available online: https://www-scopus-com.inc.bib.cnrs.fr/record/display.uri?eid=2-s2.0-51249088328&origin=resultslist&sort=plf-f&src=s&nlo=&nlr=&nls=&sid=d6272db4bba8532cdcf678074490a4cc&sot=a&sdt=cl&cluster=scopubyr,%222008%22,t&sl=42&s=(AUTHOR-NAME(zhao))+AND+(effect+Caseinate)&relpos=1&citeCnt=22&searchTerm=&featureToggles=FEATURE_NEW_DOC_DETAILS_EXPORT:1 (accessed on 26 April 2022).
22. Srikaeo, K.; Doungjan, S. Hydrocolloids Retard Lipid Digestion in Whipping Cream. *Food Res.* **2020**, *4*, 831–838. [CrossRef] [PubMed]
23. Han, Y.; Zhu, L.; Qi, X.; Zhang, H.; Wu, G. Characteristics of Low-fat Whipped Cream Containing Protein-based Fat Replacers. *Int. J. Dairy Technol.* **2023**, *76*, 276–290. [CrossRef]
24. Ghribi, A.M.; Zouari, M.; Attia, H.; Besbes, S. Study of Protein/k-Carrageenan Mixture's Effect on Low-Fat Whipping Cream Formulation. *LWT* **2021**, *147*, 111647. [CrossRef]
25. Grossi, M.; Rao, J.; Chen, B. Tuning the Properties of Plant-Based Whipped Cream through Zein Nanoparticles-Sodium Stearate Complex. *Food Hydrocoll.* **2024**, *155*, 110219. [CrossRef]
26. Cao, Z.; Liu, Z.; Zhang, H.; Wang, J.; Ren, S. Protein Particles Ameliorate the Mechanical Properties of Highly Polyunsaturated Oil-Based Whipped Cream: A Possible Mode of Action. *Food Hydrocoll.* **2020**, *99*, 105350. [CrossRef]
27. Lim, C.W.; Norziah, M.H.; Lu, H.F. Effect of Flaxseed Oil towards Physicochemical and Sensory Characteristic of Reduced Fat Ice Creams and Its Stability in Ice Creams upon Storage. *Int. Food Res. J.* **2010**, *17*, 393–403.
28. Shamsi, K.; Che Man, Y.B.; Yusoff, M.S.A.; Jinap, S. A Comparative Study of Dairy Whipping Cream and Palm Oil-based Whipping Cream in Terms of FA Composition and Foam Stability. *J. Am. Oil Chem. Soc.* **2002**, *79*, 583–588. [CrossRef]
29. Li, Y.; Liao, T.; Liu, T.; Yan, R.; Sun, Z.; Zhao, M.; Deng, X.; Zhao, Q. The Quality of Whipped Cream: Effect of Polyglycerol Ester on the Crystallization of Fat Blend and the Properties of Interface. *Food Hydrocoll.* **2023**, *145*, 109145. [CrossRef]
30. Zhao, Q.; Zhao, M.; Yang, B.; Cui, C. Effect of Xanthan Gum on the Physical Properties and Textural Characteristics of Whipped Cream. *Food Chem.* **2009**, *116*, 624–628. [CrossRef]
31. Lazzaro, F.; Saint-Jalmes, A.; Violleau, F.; Lopez, C.; Gaucher-Delmas, M.; Madec, M.N.; Beaucher, E.; Gaucheron, F. Gradual Disaggregation of the Casein Micelle Improves Its Emulsifying Capacity and Decreases the Stability of Dairy Emulsions. *Food Hydrocoll.* **2017**, *63*, 189–200. [CrossRef]
32. Allen, K.E.; Dickinson, E.; Murray, B. Acidified Sodium Caseinate Emulsion Foams Containing Liquid Fat: A Comparison with Whipped Cream. *LWT* **2006**, *39*, 225–234. [CrossRef]
33. Marinova, K.G.; Basheva, E.S.; Nenova, B.; Temelska, M.; Mirarefi, A.Y.; Campbell, B.; Ivanov, I.B. Physico-Chemical Factors Controlling the Foamability and Foam Stability of Milk Proteins: Sodium Caseinate and Whey Protein Concentrates. *Food Hydrocoll.* **2009**, *23*, 1864–1876. [CrossRef]
34. Li, M.; Li, Y.; Wang, R.; Wang, Y.; Li, Y.; Zhang, L. Effects of Triglycerol Monostearate on Physical Properties of Recombined Dairy Cream. *Int. Dairy J.* **2020**, *103*, 104622. [CrossRef]
35. Herrero, A.M.; Carmona, P.; Pintado, T.; Jiménez-Colmenero, F.; Ruíz-Capillas, C. Olive Oil-in-Water Emulsions Stabilized with Caseinate: Elucidation of Protein–Lipid Interactions by Infrared Spectroscopy. *Food Hydrocoll.* **2011**, *25*, 12–18. [CrossRef]
36. Wang, J.; Zeng, D.; Yan, R.; Huangfu, J.; Hu, Q.; Cai, Y.; Liu, T.; Zhao, M.; Zhao, Q. Investigating the Impact of Static Destabilization Mechanism on Fat Crystallization Dynamics, Emulsion Rheology, and Whipping Properties of Whipping Cream. *Food Chem.* **2025**, *463*, 141272. [CrossRef] [PubMed]
37. Liu, P.; Huang, L.; Liu, T.; Cai, Y.; Zeng, D.; Zhou, F.; Zhao, M.; Deng, X.; Zhao, Q. Whipping Properties and Stability of Whipping Cream: The Impact of Fatty Acid Composition and Crystallization Properties. *Food Chem.* **2021**, *347*, 128997. [CrossRef] [PubMed]
38. Li, Y.; Li, Y.; Li, D.; Wang, S.; Zhang, L. Progress in Research on Whipping Cream Quality and the Factors Influencing It. *Shipin Kexue/Food Sci.* **2022**, *43*, 327–335.

39. Nguyen, V.; Rimaux, T.; Truong, V.; Dewettinck, K.; Van Bockstaele, F. Fat Crystallization of Blends of Palm Oil and Anhydrous Milk Fat: A Comparison between Static and Dynamic-Crystallization. *Food Res. Int.* **2020**, *137*, 109412. [CrossRef]
40. Liao, W.; Gharsallaoui, A.; Dumas, E.; Elaïssari, A. Understanding of the Key Factors Influencing the Properties of Emulsions Stabilized by Sodium Caseinate. *Food Sci. Food Saf.* **2022**, *21*, 5291–5317. [CrossRef]
41. Zhan, Y.; Wei, L.; Ge, T.; Chen, D.; Wang, L.; Jiang, J. Enhancing Whipped Cream Anti-Freeze Properties: A Dual Plant System with Aquafaba and Mung Bean Protein. *Food Hydrocoll.* **2025**, *159*, 110700. [CrossRef]
42. Zhao, Q.; Kuang, W.; Long, Z.; Fang, M.; Liu, D.; Yang, B.; Zhao, M. Effect of Sorbitan Monostearate on the Physical Characteristics and Whipping Properties of Whipped Cream. *Food Chem.* **2013**, *141*, 1834–1840. [CrossRef]
43. Li, Y.; Liao, T.; Liu, T.; Wang, J.; Sun, Z.; Zhao, M.; Deng, X.; Zhao, Q. Effect of Stearic and Oleic Acid-Based Lipophilic Emulsifiers on the Crystallization of the Fat Blend and the Stability of Whipped Cream. *Food Chem.* **2023**, *428*, 136762. [CrossRef] [PubMed]
44. Zeng, D.; Cai, Y.; Liu, T.; Huang, L.; Wang, J.; Zhao, M.; Zhu, S.; Zhao, Q. Effect of Hydrophobic Sucrose Esters with Different Fat Acid Composition and Esterification Degree on Whipped Cream Properties. *Food Hydrocoll.* **2024**, *146*, 109183. [CrossRef]
45. Xie, P.; Ji, G.; Jin, J.; Xu, H.; Zhang, L.; Gao, Z.; Fan, Q.; Chen, Y.; Jin, Q.; Wang, X. Partial Coalescence and Whipping Capabilities of Dairy Aerated Emulsions as Affected by Inclusion of Monoglycerides with Different Fatty Acid Species. *Int. J. Dairy Technol.* **2023**, *76*, 81–92. [CrossRef]
46. Denkov, N.; Tcholakova, S.; Cholakova, D. Surface Phase Transitions in Foams and Emulsions. *Curr. Opin. Colloid Interface Sci.* **2019**, *44*, 32–47. [CrossRef]
47. Li, Y.; Li, Y.; Yuan, D.; Wang, Y.; Li, M.; Zhang, L. The Effect of Caseins on the Stability and Whipping Properties of Recombined Dairy Creams. *Int. Dairy J.* **2020**, *105*, 104658. [CrossRef]
48. Fu, L.; He, Z.; Zeng, M.; Qin, F.; Chen, J. Effects of Soy Protein Composition in Recombined Soy-based Cream on the Stability and Physical Properties of Whipping Cream. *J. Sci. Food Agric.* **2020**, *100*, 2732–2741. [CrossRef]
49. Zhang, J.; Zhu, R.; Meng, Z. Whipped Cream Stabilized by Faba Bean Protein Isolate Microgel Particles Substituted for Sodium Caseinate: Whipping Performance and Foam Stabilization Analysis. *Food Hydrocoll.* **2024**, *157*, 110388. [CrossRef]
50. Gao, K.; Rao, J.; Chen, B. Plant Protein Solubility: A Challenge or Insurmountable Obstacle. *Adv. Colloid. Interface Sci.* **2024**, *324*, 103074. [CrossRef] [PubMed]
51. Kew, B.; Holmes, M.; Stieger, M.; Sarkar, A. Review on Fat Replacement Using Protein-Based Microparticulated Powders or Microgels: A Textural Perspective. *Trends Food Sci. Technol.* **2020**, *106*, 457–468. [CrossRef] [PubMed]
52. Wei, L.; Song, K.; Shao, D.; Dong, C.; Wang, L.; Jiang, J. The Effect of Endogenous Phenol and Gallic Acid on Structure and Functionalities of Mung Bean Protein in Aerated Emulsion System. *Food Hydrocoll.* **2023**, *143*, 108942. [CrossRef]
53. Wu, S.; Zhang, Z.; Liu, C.; Ma, T. Effect of PH-Shifting and Sonication-Assisted Treatment on Properties and Stability of Vegetable Oil-Based Whipped Cream Stabilized by Kidney Bean Protein Aggregates. *Food Hydrocoll.* **2023**, *141*, 108736. [CrossRef]
54. Liu, Z.; Xu, M.; Zhou, S.; Wang, J.; Huang, Z. Enhancing the Thermal Stability of Zein Particle-Stabilized Aeratable Emulsions Through Genipin-Protein Cross-Linking and Its Possible Mechanism of Action. *J. Agric. Food Chem.* **2024**, *72*, 3707–3718 [CrossRef] [PubMed]
55. Guo, J.; Zhou, Q.; Liu, Y.-C.; Yang, X.-Q.; Wang, J.-M.; Yin, S.-W.; Qi, J.-R. Preparation of Soy Protein-Based Microgel Particles Using a Hydrogel Homogenizing Strategy and Their Interfacial Properties. *Food Hydrocoll.* **2016**, *58*, 324–334. [CrossRef]
56. Shehzad, Q.; Liu, Z.; Zuo, M.; Wang, J. The Role of Polysaccharides in Improving the Functionality of Zein Coated Nanocarriers Implications for Colloidal Stability under Environmental Stresses. *Food Chem.* **2024**, *431*, 136967. [CrossRef]
57. Kasaai, M.R. Zein and Zein -Based Nano-Materials for Food and Nutrition Applications: A Review. *Trends Food Sci. Technol.* **2018** *79*, 184–197. [CrossRef]
58. Wu, S.; Wang, C.; Liu, C.; He, Q.; Zhang, Z.; Ma, T. Synergistic Effects of Xanthan Gum and β-Cyclodextrin on Properties and Stability of Vegetable Oil-Based Whipped Cream. *Int. J. Biol. Macromol.* **2024**, *279*, 135379. [CrossRef]
59. Del Valle, E.M.M. Cyclodextrins and Their Uses: A Review. *Process Biochem.* **2004**, *39*, 1033–1046. [CrossRef]
60. Jiang, X.; Shekarforoush, E.; Muhammed, M.K.; Whitehead, K.A.; Arneborg, N.; Risbo, J. Lactic Acid Bacteria as Structural Building Blocks in Non-Fat Whipping Cream Analogues. *Food Hydrocoll.* **2023**, *135*, 108137. [CrossRef]
61. Athari, B.; Nasirpour, A.; Saeidy, S.; Esehaghbeygi, A. Physicochemical Properties of Whipped Cream Stabilized with Electrohydrodynamic Modified Cellulose. *J. Food Process Preserv.* **2021**, *45*, 15688. [CrossRef]
62. Seddiqi, H.; Oliaei, E.; Honarkar, H.; Jin, J.; Geonzon, L.C.; Bacabac, R.G.; Klein-Nulend, J. Cellulose and Its Derivatives: Towards Biomedical Applications. *Cellulose* **2021**, *28*, 1893–1931. [CrossRef]
63. Madhushree, M.; Vairavel, P.; Mahesha, G.; Bhat, K.S. A Comprehensive Review of Cellulose and Cellulose-Based Materials Extraction, Modification, and Sustainable Applications. *J. Nat. Fibers* **2024**, *21*, 2418357. [CrossRef]
64. Anukiruthika, T.; Moses, J.A.; Anandharamakrishnan, C. Electrohydrodynamic Drying of Foods: Principle, Applications, and Prospects. *J. Food Eng.* **2021**, *295*, 110449. [CrossRef]
65. Biglarian, N.; Rafe, A.; Shahidi, S. Effect of Basil Seed Gum and κ-carrageenan on the Rheological, Textural, and Structural Properties of Whipped Cream. *J. Sci. Food Agric.* **2021**, *101*, 5851–5860. [CrossRef] [PubMed]
66. Zhao, Q.; Zhao, M.; Li, J.; Yang, B.; Su, G.; Cui, C.; Jiang, Y. Effect of Hydroxypropyl Methylcellulose on the Textural and Whipping Properties of Whipped Cream. *Food Hydrocoll.* **2009**, *23*, 2168–2173. [CrossRef]

7. Cai, Y.; Zeng, D.; Huang, L.; Zhao, M.; Zhao, Q.; Van der Meeren, P. Emulsifying and Whipping Properties of Mixing Polysaccharide Dispersions: Effect of Ratio between Insoluble Soybean Fiber and Hydroxypropyl Methylcellulose. *J. Sci. Food Agric.* **2022**, *102*, 6707–6717. [CrossRef]
8. Rezvani, F.; Abbasi, H.; Nourani, M. Effects of Protein–Polysaccharide Interactions on the Physical and Textural Characteristics of Low-fat Whipped Cream. *J. Food Process Preserv.* **2020**, *44*, 14743. [CrossRef]

MDPI AG
Grosspeteranlage 5
4052 Basel
Switzerland
Tel.: +41 61 683 77 34

Molecules Editorial Office
E-mail: molecules@mdpi.com
www.mdpi.com/journal/molecules